# A Little History of the Earth

# INSPIRING GUIDES FOR CURIOUS MINDS

Whether you know absolutely nothing about a subject or are already familiar with it, these *Little Histories* are the most energetic, entertaining and reliable guides you will find.

*A Little History of the World* by E.H. Gombrich
*A Little Book of Language* by David Crystal
*A Little History of Philosophy* by Nigel Warburton
*A Little History of Science* by William Bynum
*A Little History of Literature* by John Sutherland
*A Little History of the United States* by James West Davidson
*A Little History of Religion* by Richard Holloway
*A Little History of Economics* by Niall Kishtainy
*A Little History of Archaeology* by Brian Fagan
*A Little History of Poetry* by John Carey
*A Little History of Art* by Charlotte Mullins
*A Little History of Music* by Robert Philip
*A Little History of Psychology* by Nicky Hayes
*A Little History of Mathematics* by Snezana Lawrence
*A Little History of the Earth* by Jamie Woodward

New titles coming soon!

Discover more about the full series at:
yalebooks.co.uk/little-histories   yalebooks.com/little-histories

# A Little
# History of
# The Earth

# Jamie
# Woodward

YALE UNIVERSITY PRESS
NEW HAVEN AND LONDON

For information about this and other Yale University Press publications, please contact:
U.S. Office: sales.press@yale.edu yalebooks.com
Europe Office: sales@yaleup.co.uk yalebooks.co.uk

Set in Minion Pro by IDSUK (DataConnection) Ltd

Printed and bound in the UK using 100% renewable electricity at CPI Group (UK) Ltd

Library of Congress Control Number: 2025940612
A catalogue record for this book is available from the British Library.
Authorized Representative in the EU: Easy Access System Europe, Mustamäe tee 50, 10621 Tallinn, Estonia, gpsr.requests@easproject.com

ISBN 978-0-300-24614-8

10 9 8 7 6 5 4 3 2 1

*For Danny Woodward 1963–2025*

# Contents

| Eon | Era | Period | Epoch | Onset (m.y.a.) | |
|---|---|---|---|---|---|
| PHANEROZOIC | Cenozoic | Quaternary | Holocene | 0.0117 | First farmers, urban civilisation, Industrial Revolution, Great Acceleration |
| | | | Pleistocene | 2.58 | Ice age, Neanderthal extinction, Homo sapiens go global |
| | | Neogene | Pliocene | 5.33 | North and South America join and animals migrate in both directions |
| | | | Miocene | 23.04 | Mediterranean Sea dries up |
| | | Palaeogene | Oligocene | 33.9 | Antarctica buried under ice, Alps form in Europe |
| | | | Eocene | 56.0 | Big spike in global warming, Himalayas begin to form |
| | | | Palaeocene | 66.0 | Extensive forests, mammals dominate land-based ecosystems |
| | Mesozoic | | Cretaceous | 143.1 | Asteroid strike and extinction of dinosaurs 66 million years ago |
| | | | Jurassic | 201.4 | Dinosaurs and marine reptiles dominate Earth's ecosystems |
| | | | Triassic | 251.9 | Emergence of mammals and first dinosaurs on Pangaea |
| | Palaeozoic | | Permian | 298.9 | Great Dying – the largest mass extinction at close of Permian |
| | | | Carboniferous | 358.86 | Supercontinent Pangaea forms, swampy forests produce coal deposits |
| | | | Devonian | 419.62 | The Age of Fishes |
| | | | Silurian | 443.1 | Plant cover increases on land and coral reefs expand in the oceans |
| | | | Ordovician | 486.85 | Ice sheets form on a giant southern landmass, mass extinction in the ocean |
| | | | Cambrian | 538.8 | Burst of new animals on the ocean floor |
| PROTEROZOIC | Neoproterozoic | | Ediacaran | 635.0 | First appearance of soft-bodied animals in the oceans |
| | | | Cryogenian | 720.0 | Snowball Earth periods – global freezes followed by rapid warming |
| | | | Tonian | 1000.0 | Huge landmass forms in the southern hemisphere |
| | Mesoproterozoic | | Stenian | 1200.0 | Widespread mountain formation, tiny algae still dominate life in the oceans |
| | | | Ectasian | 1400.0 | Supercontinent Rodinia begins to form |
| | | | Calymmian | 1600.0 | Area of land increases and first fungi in the geological record |
| | Palaeoproterozoic | | Statherian | 1800.0 | Oxygen content of the atmosphere continues to rise |
| | | | Orosirian | 2050.0 | Earliest molecular evidence for sexual reproduction |
| | | | Rhyacian | 2300.0 | Big ice age and global cooling |
| | | | Siderian | 2500.0 | Iron-rich rocks laid down in the oceans |
| ARCHEAN | Neoarchean | | | 2800.0 | Oxygen content of the atmosphere rises |
| | Mesoarchean | | | 3200.0 | Plate tectonics in operation |
| | Palaeoarchean | | | 3600.0 | Earliest undisputed evidence of life |
| | Eoarchean | | | 4031.0 | Oldest known rocks, Earth bombarded by comets and meteorites |
| HADEAN | | | | 4567.0 | Formation of the Earth and Moon, early atmosphere and oceans |

Precambrian

The ages for geological periods, epochs etc. in this book are from the International Commission on Stratigraphy (https://stratigraphy.org/). In most cases they are rounded to the nearest 0.1 million years. Billions of years are also used for events in the Precambrian. The big five mass extinctions are shown. Note that m.y.a. = million years ago.

# CHAPTER 1

# A View into Deep Time

Each spring, with a group of colleagues and undergraduate students, I take part in a field course in the High Atlas Mountains of Morocco. It is one of my favourite places to explore Earth history and landscape change. We stay in Imlil, a bustling village and a base for trekkers heading to Jebel Toubkal, the highest mountain in North Africa. The setting is spectacular: there are towering snow-capped peaks, sheer basalt cliffs, and bedrock channels fed by cascading waterfalls that disappear into terraced valley-bottoms shaded by fruit trees in blossom. Enormous boulders from a catastrophic avalanche scar the slopes around the village.

We can see Earth's history written around us. This landscape tells a story of ancient lava flows, supercontinent breakup, tropical river flooding, tectonic uplift, devastating earthquakes and mountains on the fringe of the Sahara scoured by ice age glaciers. All of this and more challenges us to think about deep time, the changing geography of the globe, shifting climates and geohazards. We also think about how the geological record is formed and why, in some areas, large parts of the record are missing. We consider human impacts on this mountain landscape too, since trees were first felled and goats began grazing. Did these processes trigger soil erosion in historical times? What does the warming climate of today's Earth mean for these mountain ecosystems, for snowpack survival and water availability? All of this generates lively conversations about Earth as a system, now and in the past.

The Earth is a system because it is made up of components that interact. These components exchange energy and matter at scales from global to local. We can think about these components as the five great spheres of the Earth system: the *lithosphere* (the solid Earth including the crust, mantle and core), *hydrosphere* (liquid water), *cryosphere* (frozen water), *biosphere* (all living things, including humans and organic matter, and all the places where life exists) and *atmosphere* (the gases that surround the Earth). Earth is essentially a closed system for matter (almost nothing gets in or out), but an open system for energy because we receive light and heat from the Sun and radiate heat into space. When one part of the Earth system changes the others respond, which may involve various feedbacks – when a change in one part of the system causes effects that either amplify (positive feedback) or dampen (negative feedback) the original change. The responses may be rapid or exceedingly slow. This book tells the story of how the elements of the Earth system have evolved and interacted since the earliest times to create a dynamic, interconnected whole – and how humans are now a key participant in this system.

There are three main rock types around Imlil. The 4,167-metre-high Toubkal and its neighbouring peaks are formed in massive, layered stacks of dark basalt. These rocks hardened from lavas which flowed across the Earth's surface during the breakup of a supercontinent. The lava flows were generated by intense volcanic activity in the latter part of a geological period known as the Ediacaran, which spanned 635 to 538.8 million years ago. It was during the Ediacaran that the first complex animals emerged in the ocean.

The barren, soil-stripped lower slopes around Imlil are dominated by a hard crystalline rock called diorite. It is older than the basalt (the first lavas flowed across its surface) but also dates to the Ediacaran Period, when it solidified in magma chambers deep below the Earth's surface. In one valley we see beds of pink sandstone and siltstone that were laid down much later in the Triassic Period (251.9 to 201.4 million years ago), when the first dinosaurs appeared. These sediments, stained by iron oxide in a sweltering greenhouse climate, were deposited by a large tropical river (rather like today's Nile) that saw big wet season floods. A large chunk of the geological record is missing – in this part of the High Atlas Mountains there is a gap of about 300 million years between the diorite and sandstones.

The High Atlas Mountains extend for some 750 kilometres across northwest Africa from the Atlantic coast of Morocco in the west to the border with Algeria in the east. Over the last 50 million years or so, during the Cenozoic Era, these mountains have been squeezed upwards by the collision of the African and Eurasian tectonic plates. Tectonic plates are huge pieces of Earth's outer shell that fit together like a giant jigsaw puzzle and slowly move over time. Immense forces are generated where tectonic plates meet. On 8 September 2023, a major earthquake struck this region, causing widespread destruction. Thousands of people lost their lives.

The High Atlas is an extraordinary place that allows us to peer into these periods in Earth history. We examine the local geomorphology too – this is the study of the landforms at the Earth's surface and the processes that shape them. The High Atlas saw glaciers develop in at least three periods of cold climate during the Pleistocene Epoch (2.58 million years to 11,700 years ago), when summers were cooler and thick snowpacks persisted year-round. On the northern slopes of the Toubkal Massif above 2,000 metres, there are beautifully preserved moraines – as clear as any you might see in the North American Rockies or the Swiss Alps – with sharp ridges and glacially scratched basalt boulders. The most extensive glaciation in the High Atlas reached its maximum extent about 50,000 years ago.

Towards the end of the eighteenth century, the Scottish geologist James Hutton (1726–1797) wrote down his thoughts about deep time. He argued that geological processes operate slowly and continuously over immense spans of time, challenging the established belief in a young Earth taken from the Book of Genesis. In the twentieth century the advent of scientific dating methods – using elements with radioactive isotopes as geological clocks – showed that the Earth was about 4.54 billion years old, thereby providing a timescale for deep time and Earth history. The Toubkal basalts solidified about 0.6 billion years ago. Thinking about geological time is difficult because it is so vast compared with our own lived experience.

The geological timescale is one of the greatest achievements in the history of science. It is the result of over two hundred years of painstaking work piecing together parcels of information from outcrops and fossil beds scattered around the world. The Harvard palaeontologist and evolutionary biologist Stephen Jay Gould (1941–2002) viewed the geological timescale as a brilliant synthesis of fossil evidence, stratigraphy (the study of rock layers and their ordering) and relative dating, constructed through careful observation and reasoning. It stands alongside the periodic table for chemistry as one of the greatest frameworks in science. It provides the backbone for understanding Earth's history, the evolution of life, as well as billions of years of climate change. It also provides a structure for understanding the behaviour of Earth as a system. You can find a diagram showing the timescale at the beginning of this book.

In 1990, when I was a graduate student in Cambridge, England, I attended a lecture by Stephen Jay Gould. His subject was deep time and the nature of evolution. One of Gould's favourite ways of picturing deep time was by comparing Earth history to the length of an outstretched arm from shoulder to fingertip. If the shoulder marks the formation of the Earth (4.54 billion years ago), the elbow marks the rapid build-up of oxygen (about 2.5 billion years ago) alongside some of the first simple lifeforms in the oceans, with the age of the dinosaurs taking place within a couple of centimetres on the palm of your hand. Finally, the age of humans begins with the last millimetre of the fingernail on the middle finger. Earth and life have a history that spans billions of years; this analogy places the brevity of human existence into sharp perspective against the vast history of the planet. A single stroke of a nail file erases all of recorded human history.

Gould wrote extensively about how profoundly the discovery of deep time influenced human thought. By highlighting how brief and recent our presence on Earth actually is, the reality of deep time forced us to reconsider human importance. The Earth was not created for us as the Old Testament would have us believe. If the geological timescale reframed our place in nature, another radical reframing has been mooted since 2000. A decade after that lecture in Cambridge an influential group of environmental scientists began to use the term Anthropocene to describe the time that we inhabit now – a compelling case has been made for formally incorporating an Anthropocene into the geological timescale.

Although human history is but a tiny fraction of Earth history, our impacts – through industry, urban development, agriculture, deforestation, pollution, climate change and biodiversity loss – have accelerated so rapidly since the mid-twentieth century, and influenced the Earth system so profoundly, that many argue we have entered a new chapter in the planet's geological story.

In February 2025 I attended a lecture on global climate change at the Royal Geographical Society in London by Ed Hawkins from Reading University. The title of the lecture was 'Why 1°C Matters'. Ed Hawkins devised the now iconic climate stripes visualisation that has started conversations about climate change in all parts of the world, encouraging everyday reflections on global warming and what the future may bring. The graphic is elegant and simple: a pattern of coloured stripes showing changing temperatures since 1850 – one stripe for each year in this climate history – with dark blues for colder years and dark reds for warmer years. From a single glance one can see the inexorable march towards rapid warming. The climate stripes form the artwork on the cover of Greta Thunberg's 2022 *The Climate Book* and have appeared on the covers of newspapers and magazines around the world. They have even been projected onto the White Cliffs of Dover. In many respects, the rapid rise in atmospheric carbon dioxide concentrations, which has driven the recent warming of our planet, is the defining feature of the Anthropocene.

In her 2018 book *Timefulness: How Thinking Like a Geologist Can Help Save the World*, the American geologist Marcia Bjornerud bemoans the fact that few people fully comprehend the vast time-scale of Earth history. She argues that such a time-limited perspective underpins many of the environmental problems we now face. Bjornerud explains why *timefulness* – a deeper understanding of geological time and the very ancient rhythms of Earth history – is of critical importance if we are to move away from short-term thinking and navigate an Anthropocene of runaway planetary change. People need to grasp how rapidly the Earth is now changing.

This is a book of Earth stories that examine the great span of Earth history, from the very beginnings and evolution of the planet and its lifeforms to the present day. Its aim is not just to give context to current environmental concerns, but to explore questions you might never have even asked about the world you see around you: how it came to be and how it works, and what we can do to preserve it. How did a disc of hot space dust become a habitable planet teeming with life? Where did all the water on this blue planet come from? When did plate tectonics begin? When did life emerge in the oceans and on land? What caused the big five mass extinctions and are we living in the sixth? These are just some of the questions we will explore. All this, of course, demands an engagement with deep time and shifts in global geography. So first we need to turn back the clock to sometime before 4.54 billion years ago. It's time to start building a planet.

# CHAPTER 2

# How to Build a Planet

To understand the birth of planet Earth and our Solar System – the Sun, eight planets, hundreds of moons, dwarf planets, comets, asteroids, and lumps of ice and rock too numerous to count – we must first think about the deaths of stars.

When some stars reach the end of their lives they collapse with a catastrophic explosion that blasts gases and dust far across space. This matter is the raw material for new stars and planets. But where did this material come from?

About 13.8 billion years ago the universe started from a tiny point of infinite density and temperature and suddenly began to expand rapidly. This is the Big Bang, marking the beginning of space and time. The universe cooled as it expanded, allowing particles to form atoms of mainly hydrogen and helium. And this led, over millions of years, to the formation of stars and galaxies.

There are somewhere between 100 and 400 billion stars in our galaxy and about 2 trillion galaxies in the observable universe – the part of the universe we can see from Earth, which is limited by the speed of light. Stars are the workshops of the universe: the nuclear reactions that take place in the searing heat of solar cores convert lighter elements into heavier ones – the specific elements a star produces depend on its size and life-cycle stage. Temperatures can exceed 15 million degrees Celsius in the core of our nearest star, the Sun. This unimaginable heat is fuelled by the nuclear conversion of hydrogen atoms to helium atoms under intense pressure. As a star reaches the end of its life, it may produce heavier elements such as oxygen, carbon, nitrogen and iron. Stars of all sizes create helium, and more massive stars create heavier elements and disperse them through the universe. This has been the order of things since the Big Bang, leaving about 9 billion years to make the heavier elements in our Solar System.

Our cosmic ancestry is mind-blowing. Everything on Earth, including you and me, is ultimately derived from the exploded remnants of dying stars – from dust and gas that sailed across space.

Stars are so massive they generate immensely powerful gravitational forces that try to compress their mass inwards into the smallest size possible. In opposition to this gravity is the heat generated in the core of the star by nuclear fusion – this heat drives expansion and forces matter outwards, balancing the force of gravity. When the core of a star finally runs out of fuel and its hydrogen stock approaches exhaustion, it collapses and gravity wins the day. The end is sudden and violent. A star can collapse in about ten to twenty seconds, generating a gigantic explosion. This is a supernova.

Our Solar System began to form about 4.6 billion years ago within a dense cloud of interstellar dust and hydrogen gas. One theory suggests this cloud began to collapse when it was hit by the shockwave from an exploding star – the dust cloud began to rapidly contract under the force of its own gravity. As the core became denser, the heat and pressure became so intense that nuclear fusion could take place in its centre and hydrogen began converting to helium. When hydrogen atoms fuse together into helium, the reaction releases enough energy to balance the gravitational forces trying to collapse the gas cloud. These forces led to the formation of our Sun. The remaining dust and gas in the interstellar cloud formed a rotating disc that astronomers call the solar nebula.

In the inner Solar System, solid rocky and metallic material that could withstand the heat of the young Sun began to clump together in the solar nebula to form the terrestrial planets with hard rocky surfaces (Mercury, Venus, Earth and Mars). Further away from the Sun, there was more solid material that built planetary cores big enough to capture surrounding gas, leading to the formation of Jupiter and Saturn (the gas giants) and, finally, where the warmth of the Sun was much diminished, solid icy material aggregated to form two giant ice planets, Uranus and Neptune.

The law of universal gravitation formulated by the great English polymath Isaac Newton (1642–1727) states that every particle attracts every other particle in the universe with a force that varies according to their mass and the distance between them. The movement of particles in the solar nebula was controlled by the Sun's gravity. The largest objects have the strongest gravity, and this force weakens with distance from the object. The Sun accounts for over 99 per cent of all the mass in our Solar System, with all the planets, moons, comets, asteroids and other interstellar matter making up the rest. Because the Sun is so big (over 330,000 times the mass of Earth) and its gravitational pull so dominant, it literally holds the Solar System together, keeping all the planets in orbit around it. It has enough fuel to keep shining for at least another 5 billion years.

The young Sun was surrounded by the 1 per cent of material left in the solar nebula. Particles of matter in this disc began to collide and stick together in a process called accretion. While the forces involved are not fully understood, this is the attraction and accumulation of particles into larger masses that can grow into a planet-sized body.

Millions of larger objects called planetesimals were formed in a violent and chaotic process. These planetesimals further accreted to form even larger bodies known as protoplanets, which eventually collided to become the planets of the Solar System. Earth emerged from trillions of particle collisions forced together by gravity, while Mercury may be a relic of a huge collision, and our Moon is a product of one.

Accretion happens at all scales, from the tiniest cosmic dust particle to planet-sized bodies. With powerful radiotelescopes it is now possible to observe accretion taking place as new solar systems form in distant galaxies. And it is still happening in our Solar System today as asteroids collide and meteorites fall to Earth.

The next time you visit a museum with a rock collection, go and have a look at the meteorites on display. Some meteorites are samples of the original building blocks of planets, lumps of rock from a time before Earth history. Some are remnants of the solar nebula that have avoided planetary collisions for 4.6 billion years. If you touch a meteorite, you could be touching an object that is the same age as the Sun.

Most meteorites are samples of asteroids and there are two main types. Primitive meteorites have changed very little since the time of the early Solar System – they were flying through space at the time when Earth was forming. The second group is called processed meteorites. These have been involved in the formation of a planet or moon and have been altered by that process. They may have been blasted from that body by a collision and have properties that are similar to the crust or mantle or core of a planet. Some asteroids are fragments of the cores of planetesimals. Processed meteorites tend to be younger than their primitive cousins.

Our Solar System is still full of dust and rocks. A lump of rock flying through space is called a meteoroid. When it enters Earth's atmosphere at high speed it starts to burn up and glow brightly. This is a meteor, commonly known as a shooting star. It has been estimated that about 44 tonnes of meteoroid enter the Earth's atmosphere every day. Most of this space rock burns up in the atmosphere, never to reach the ground. A meteor only becomes a meteorite when some of the rock lands on the Earth's surface. About 6,000 meteorites fall to the Earth's surface each year. Most are never recovered but, according to the USA's National Aeronautics and Space Administration (NASA), over 50,000 meteorites have been found on Earth. Many of these originated in the asteroid belt that lies between Mars and Jupiter.

Planetary scientists get very excited when a new meteorite arrives that can be recovered and studied. Detailed study of meteorites has provided valuable information about the chemical conditions in the solar nebula, yielding clues about the processes involved in the formation of our world. The largest meteorite ever discovered on Earth was found by a farmer just below the surface of a field in Namibia in 1920. The Hoba meteorite weighs about 66 tonnes and is composed of iron (84 per cent) and nickel (16 per cent). It is so large and heavy it has never been moved to a museum – a visitor display was constructed around it and thousands of people visit every year.

The young Earth was a hot and molten mass due to the energy released during the accretion process. The collisions that took place during Earth's formation generated so much heat that the young planet's surface was dominated by a vast ocean of boiling magma. The surface was incandescently hot with heat added by frequent bombardment by asteroids and comets. We've all seen lava flows on TV, but imagine an ocean basin of bubbling magma all the way to the horizon releasing hot gases into the atmosphere. The magma ocean existed in the early part of what geologists call, after the ancient Greek underworld, the Hadean Eon (4.567 to 4.031 billion years ago). It must have been a truly hellish place. When parts of this lava ocean cooled it formed the first rocky crust. Most of the rocks from this period formed over 4 billion years ago and have been eroded away or recycled by plate tectonics, and so direct geological evidence of these magma oceans is not easy to find. The onset of the Hadean Eon at 4.567 billion years ago comes from scientific dating of meteorites that represent the oldest known solid material in the Solar System.

The Earth's core, mantle and crust (Chapter 8) began to form in the Hadean. In the early molten state, chemical elements built up in different zones of the emerging planet according to their density and other properties. The heaviest materials, such as iron and nickel, sank deep into the planet to form the core while lower-density molten material rose to the surface to form the crust. This process is called differentiation. Think of iron ore breaking down in the searing heat of a furnace; the lighter impurities rise to the surface to form a crust, known as slag, above the much denser molten iron. This is a form of differentiation like the planetary-scale process that led to a dense iron core and a lighter rocky crust on Earth. As Earth's first crust solidified, heat transfer from the interior to the surface was much reduced, so the atmosphere cooled enough to allow water to condense.

The end of the Hadean Eon also witnessed the onset of a chaotic period in the early history of our Solar System called the Late Heavy Bombardment (LHB), between about 4.1 and 3.8 billion years ago, when Earth and other planets were pummelled by a huge number of asteroids and comets. Since that time Earth has broadly kept the same amount of matter, recycling this material through systems like the water, carbon and rock cycles.

When the Solar System was forming, Earth was too hot for water to survive on its surface. So how did a blue planet with diverse ecosystems on land and in the oceans emerge from the inhospitable chaos of the early Hadean and the LHB? We will explore all this in later chapters. One crucial aspect is Earth's location in the Solar System at an average distance of 149.6 million kilometres from the Sun. This position in the habitable zone allowed Earth to become our home.

# CHAPTER 3

# The Goldilocks Zone

The first photograph to capture all of planet Earth was taken on 7 December 1972 by the crew of *Apollo 17* – the last NASA mission to the Moon. In the photo, Earth appears as a small blue sphere with swirls of white, glistening amidst the bleak, black vastness of space. The *Blue Marble*, as the photograph came to be known, has become one of the defining images of the twentieth century and one of the most viewed and reproduced photographic images of all time. It not only captured the beauty and complexity of our Earth, but also very quickly became a symbol of humanity's place in the cosmos and a powerful reminder of the fragility of our home.

The *Blue Marble* photograph was made possible because this was the first Apollo trajectory to allow a whole view of the Earth fully illuminated by the Sun. No human since has travelled far enough to capture such a whole-Earth image. But remarkably, the photograph was not intended to be taken at all. It eventually emerged that it was *Apollo 17*'s Harrison Schmitt (b. 1935) – the only geologist to have walked on the Moon – who couldn't resist snapping the full outline of our world as it came into view, despite being assigned to another task at the time.

When the photograph was taken, the *Apollo 17* spacecraft was about 29,000 kilometres from the Earth's surface. This was far enough away to see the whole planet, but close enough to see details of Earth's physical geography. In the photo we can see a living, breathing and above all watery world – from the frozen wastelands of Antarctica to the Atlantic, Indian and Southern Oceans, to the white swirls of vapour forming the clouds of weather systems and storms. All this involves huge volumes of water as solid, liquid and gas. The Cape of Good Hope on the southern tip of Africa – the continent where humanity first emerged – is hidden beneath the spiralling clouds of a cyclone. Look very closely and you can even trace the River Nile threading its way across the eastern Sahara transporting rainwater from the green wet tropics of Equatorial Africa to the Mediterranean Sea. All of this makes our planet unique. Earth is the only place in our Solar System with an active hydrological cycle. It is the only planet we know where water is abundant in all its forms and where water is continuously cycled between oceans, atmosphere and continents. Over 70 per cent of the surface is liquid water: Earth really is the blue planet.

There are billions of stars in the Milky Way, but how common are planets that support liquid water? Could there be planets in other solar systems with *oceans* of water? Is there another planet like Earth? Around every star is a zone of space where the temperature on a planetary surface would allow water to exist as liquid. If a planet is too close to its star, conditions are too hot and water would boil. Too far away from the warmth of the star (beyond the snowline) and water freezes. Astronomers call the range of distance where water could exist as a liquid the *habitable zone*. It is also known as the *Goldilocks zone* because conditions are neither too hot nor too cold; the distance from a sun is just right. Every star has a habitable zone, but the extent and location of that zone will vary for stars of different sizes and brightness.

Water is an essential ingredient for life (as we know it) and liquid water hosts most of the life on Earth. Rainforests in the tropics are

Earth's richest ecosystems while desert landscapes are the most impoverished. Life is patchy and challenging where liquid water is scarce. To nurture and sustain complex life, lots of water is needed. Earth sits in the centre of our Solar System's habitable zone. The Celsius temperature scale is derived from the freezing (0°C) and boiling (100°C) points of water. The average temperature on the surface of Earth is about 15°C. Temperatures are far cooler in the polar regions, of course, where huge volumes of water are frozen in ice sheets and glaciers, but temperatures across most of the Earth's surface allow water to exist as a liquid in a wide range of environments including lakes, bogs, rivers and the oceans. All these places teem with life.

The *Blue Marble* photograph encapsulates the essence of Earth as a complex, interconnected system and reminds us that all parts of Earth – atmosphere, hydrosphere, cryosphere, biosphere and lithosphere – interact to shape the environment and maintain the planet's habitability. The habitable zone in our Solar System includes Venus, Earth and Mars, but atmospheric conditions vary so profoundly that liquid water is only abundant on one of these planets.

Venus is the nearest planet to Earth and about the same size. It is the hottest planet in our Solar System: surface temperatures can exceed 475°C because of *extreme* greenhouse warming. Its super-dense atmosphere is dominated by carbon dioxide (>96 per cent) with clouds of sulphuric acid. Venus is far hotter than Mercury even though Mercury is closer to the Sun. NASA has flown probes through the atmosphere of Venus to collect data on its composition. The water content of Venusian clouds is too low to support even the most drought-tolerant microbes known on Earth. The complete absence of liquid water on the planet's surface is a fundamental barrier to life as we know it.

In stark contrast, Mars has quite large bodies of water on its surface, but all of this water is frozen. Large ice caps exist in the polar regions, but they are neither growing nor shrinking because Mars no longer has an active water cycle. The Martian landscape has deep canyons and well-preserved river channels showing that large volumes of water once flowed freely on its surface. But all of this took place billions of years ago. The Martian atmosphere is now too thin and there is no greenhouse effect to warm the planet so that it might support an active hydrological cycle. Mars may have supported life in the distant past, but there is no evidence of life today.

If life exists beyond our Solar System, it is most likely to be on planets that lie in the habitable zone of a star. Just a few decades ago, exoplanets (planets that orbit a star outside our Solar System) were the stuff of science fiction because astronomers had never identified one. All of that was to change dramatically when NASA launched the Kepler mission in 2009 to search for Earth-sized planets in the habitable zone of distant stars. This mission was named for the brilliant German astronomer Johannes Kepler (1571–1630), who described three major laws of planetary motion.

For almost a decade, between March 2009 and November 2018, the Kepler space telescope monitored the brightness of more than half a million stars in the Milky Way galaxy. When a planet is observed to pass in front of its host star, the event is called a *transit*. Each transit blocks a fraction of star light and the star's brightness dims very slightly. Imagine standing on a beach in Morocco at midnight looking to Europe across the Straits of Gibraltar. On a clear evening you can see the flickering light from the Europa Point Lighthouse. Detecting exoplanets is a bit like being able to sense the dip in brightness from the lighthouse as a mosquito flies in front of the bulb. These tiny dips in brightness were measured by the Kepler telescope. In 2014, amid huge excitement, NASA announced the discovery of exoplanet Kepler-186f, the first Earth-size planet shown to orbit the habitable zone of a star in another solar system. This exoplanet may have seasons and a stable climate. The Kepler mission has been prolific, identifying thousands of exoplanets. While we will only ever scratch the surface of the history of those planets, data from the Kepler telescope have completely trans-formed our understanding of our place in the Universe.

The most Earth-like planet so far discovered is Kepler-452b. It orbits a star known as Kepler-452 in the Cygnus constellation, some 1,800 light years from Earth. Its discovery was enormously significant because it ticks many of the boxes that suggest it may have supported life. This exoplanet is often called 'Earth 2.0', or 'Earth's Cousin'. Kepler-452b is believed to be a rocky planet with a mass five times that of Earth. It lies in the habitable zone of its star at about the same distance as Earth from the Sun. It has an orbit of 385 days. Its radius is about 60 per cent larger than Earth's with about twice as much surface gravity, so that a human would feel twice as heavy on the surface – like carrying a backpack of your own bodyweight. As soon as Kepler-452b was discovered and its similarities with Earth were set out, there was speculation about its demise. Its star is getting brighter and growing old. The study of exoplanets is bittersweet. It provides tantalising glimpses of distant worlds that may support life, but it has also revealed solar systems at different stages of their evolution. Some of them are dying.

The closest star to our Solar System is Proxima Centauri, a mere 4.2 light years from Earth, or about 265,000 times the distance between the Sun and Earth. Three exoplanets (b, c and d) have been identified orbiting this red dwarf star. The last of these, Proxima Centauri d, was confirmed early in 2022 from observations made by the world's most powerful terrestrial telescope that sits on a mountaintop in Chile. Astronomers have been able to observe how this star wobbles under the gravitational pull of its orbiting planets. Proxima d is just a quarter of the mass of Earth and one of the smallest exoplanets so far detected. This exoplanet was causing its star to wobble back and forth by 40 centimetres every second. It is staggering to contemplate that this tiny wobble can be measured by astronomers in Chile over 39 trillion kilometres away. Proxima d orbits its star every 5.12 days and is therefore too close to be in the habitable zone. While Proxima c is too cold to be habitable, Proxima b could be the closest exoplanet to Earth orbiting in the habitable zone of its star.

Identifying exoplanets is one thing; assessing exoplanet *habitability* is quite another. Will we find another Blue Marble? What we know about Venus and Mars tells us that we also need information on the nature of the atmosphere on distant planets before we can declare them water-rich and habitable. The James Webb Space Telescope launched on Christmas Day 2021 is designed to do just that. It will make transit observations in different wavelengths allowing exoplanet atmospheres to be studied in unprecedented detail. This is the new frontier in exoplanetary science.

During its operational lifetime, the Kepler telescope collected vast amounts of data as it scanned deep space for exoplanets. It is a remarkable technical achievement. NASA scientists are still exploring the files and using machine learning methods in the search for the most Earth-like planets outside our Solar System. Since 2000 the number of confirmed exoplanets has doubled about every two years. There could be billions of stars in our galaxy with between one and three exoplanets in their habitable zones. We haven't yet found an Earth-twin, but the search continues for a distant star with its own Blue Marble.

It is hard to resist being swept along by the spectacular advances of exoplanetary science, but humans will never visit Kepler-452b – it would take about 30 million years to travel there. Earth is the only habitable planet available to humanity.

The photographs taken by the crewed Apollo missions between 1968 and 1972 retain enormous significance. Harrison Schmitt's *Blue Marble* was preceded by the famous *Earthrise* photograph taken from lunar orbit on Christmas Eve 1968 by William Anders (1933–2024) aboard *Apollo 8* – a partial image of the Earth 'rising' like the Sun. Such imagery was so powerful that NASA marked the fiftieth anniversary of *Blue Marble* by taking a new whole-Earth photograph from its Deep Space Climate Observatory some 1.5 million kilometres away. We can see key changes in this new image – a reminder of the rapid pace of human impacts across the planet. The words William Anders uttered upon his return from his mission are haunting: 'We came all this way to explore the Moon, and the most important thing is that we discovered the Earth.'

# CHAPTER 4

# The Moon

There is a team at NASA that counts comets, asteroids and moons. They scan our Solar System, keeping a tally on the largest natural bodies in space, and publish the latest scientific findings. Their database lists 290 moons for our Solar System alone. Saturn has at least 146, Jupiter has ninety-five. Uranus has twenty-seven and Neptune fourteen. Even the dwarf planet Pluto has five. Mars has just two. Earth has one. Mercury and Venus have none.

The Moon is our nearest neighbour – the brightest and most recognisable object in the night sky and a source of wonder since humans first walked the Earth. Sending astronauts to the Moon – with *Apollo 11*'s Neil Armstrong (1930–2021) taking 'one giant leap for mankind' on 20 July 1969 – remains one of the greatest technological achievements of human history. The Moon is still the only place we have visited beyond our own planet. The six crewed Apollo Moon landings between 1969 and 1972 made scientific observations and collected 382 kilograms of rock and dust from the lunar surface. These samples are still being intensively studied many decades after they were returned to Earth. These samples are Earth's history too.

Moon rocks have provided vital information about the origins of Earth and the wider Solar System. The Moon formed in the aftermath of a catastrophic collision between Earth and another planet about the size of Mars. This impact sent a huge amount of molten rock and debris into space which formed a hot, rotating disc around Earth. Gravitational forces pulled this material inwards and it rapidly accreted, forming the Moon. This is known as the giant-impact hypothesis. The age of the Moon's first crust is a topic of active debate – there is some evidence pointing to its formation occurring between 4.5 and 4.4 billion years ago as the lunar magma ocean cooled and solidified.

The object that struck Earth is often called Theia, after the mythical Greek Titan and daughter of Earth goddess Gaia. Some scientists have suggested that Theia did not completely disintegrate on impact and two continent-sized slabs of Theia may lie deep in the Earth's mantle close to the core–mantle boundary. Super-computer simulations have suggested the Moon could have formed in just a few hours or days after the Theia–Earth collision, but earlier theories suggest the debris took hundreds to thousands of years to cool and accrete into a single body.

The study of the rocks and landscapes on the Moon is important because it helps to fill in key details about the early history of our world that are missing on Earth. Our Moon is a desolate and largely geologically inert environment, so its rocky crust is ancient and well preserved. Careful study of the minerals in lunar rocks has shown that before its crust solidified, the Moon's surface – like Earth's – was a deep ocean of boiling liquid magma for tens of millions of years.

Between about 4.1 and 3.8 billion years ago, during the Late Heavy Bombardment, the Moon and the Earth were still very hazardous places because they were frequently struck by comets and asteroids. The impact scars from this period can still be seen on the Moon, but they have not been preserved on the Earth's surface because they have been buried or destroyed by erosion. The surface of the Moon is heavily pocked with craters of all sizes and strewn with debris thrown around by these impact events.

The Moon's dramatic and complicated geological history has produced highlands and lowlands. From Earth we can see the dark lowland plains of basalt known as the maria. Early astronomers thought they were oceans and named them *maria*, or *mare*, which is Latin for 'seas'. The lava flows that created the *mare* were generated by partial melting of the Moon's mantle. It was once thought that lunar lava flows were triggered by massive impact events, but this theory has been abandoned because the ages of the craters and *mare* events do not match up. Before the lunar landings planetary geologists thought these maria landscapes were relatively young because they contained so few craters. This turned out not to be the case – these basalt plains are very ancient. Most volcanic activity on the Moon ended around 3 billion years ago, but some small eruptions may have continued until about 1.2 billion years ago.

*Mare Tranquillitatis*, or the Sea of Tranquillity, was the landing site for *Apollo 11* on 20 July 1969. This is where Neil Armstrong and Buzz Aldrin (b. 1930) walked on the Moon. It is one of the largest maria and can be seen with the naked eye when it comes into view before the first quarter Moon. The absence of large craters on the Sea of Tranquillity is explained by the fact that these basalts formed just after the Late Heavy Bombardment.

There is little or no erosion on the Moon's surface because it lacks an atmosphere, wind and flowing water, and has no tectonic activity. The Moon is much smaller than Earth, so it cooled more quickly after formation. This means it lacks the internal heat which drives the movement of tectonic plates on Earth (Chapter 9). The lunar crust is too thick and brittle to form tectonic plates. Since the Moon is, in key respects, geologically dead, the lunar surface changes exceedingly slowly and its impact craters are very well preserved. The age of the largest craters can be estimated by the number of smaller craters inside. The most recent analysis has logged more than 2 million craters on the surface; more than 1.3 million of these exceed 1 kilometre in diameter. There are 6,972 impact craters with a diameter of 20 kilometres or larger. The really big ones, over 150 kilometres across, are called impact basins.

The Moon has two faces, but we only see one from Earth because the Moon spins on its axis at the same rate that it orbits the Earth. This planetary choreography ensures that the same side (hemisphere) of the Moon always faces our planet. The Moon is so familiar because we always see the same features. This phenomenon, when the rotational period of an object is the same as its orbital period around another object, is known as tidal locking. This state emerges because of gravitational interaction between large bodies. Because the Moon is large and close to Earth, there is significant gravitational pull in both directions. These forces have slowed the rotation speed of both Earth and Moon. Soon after the Moon's formation, these forces synchronised the Moon's spin with its orbital period. All large moons in the Solar System are tidally locked with their planets.

The average distance to the Moon from Earth is 382,500 kilometres. This distance is not constant because the Moon moves in an elliptical orbit around the Earth and is actually moving away from us very slowly at a rate of 3.8 centimetres every year. We know this because the Apollo missions placed mirrors on the lunar surface. By directing lasers at these targets we can measure the distance with great accuracy.

In October 1959 the Soviet *Luna 3* spacecraft was the first to take images of the far side of the Moon. The images were grainy, but they revealed a heavily scarred and cratered landscape that was different to the dark lava seas of the familiar Earth-facing side. The poor quality of the images added to the mystery about the side we could not see. The crew of the *Apollo 8* mission were the first humans to leave Earth's orbit and the first to travel around the Moon. In late December 1968, they became the first to see the dark side of the Moon. Five years later Pink Floyd released their iconic album of that name. The Moon does not really have a dark side; it simply has a far side that we cannot see from Earth.

China is actively developing its lunar exploration and research using robotic vehicles and is the first nation to have visited the far side of the Moon. On 3 January 2019, China's *Chang'e 4* spacecraft landed in the Von Kármán Crater that sits inside the huge far-side crater known as the South Pole–Aitken Basin, which is some 2,500 kilometres across and 12 kilometres deep. It is the largest and deepest impact crater known in the Solar System.

The Moon has a profound impact on life on Earth. Many nocturnal creatures navigate and hunt for food by moonlight. A study in Switzerland concluded that red barn owls were less successful at hunting on moonlit nights and rodents detected their presence more easily when the Moon was full. Animal behaviourists observed various species during a total solar eclipse in 2017 as it passed over a zoo in South Carolina. Many animals mimicked nighttime behaviours as the skies darkened, while others exhibited behavioural responses associated with anxiety, including swaying and huddling.

The Moon has also inspired countless works of art. Baptised in 1564, the same year that Galileo Galilei – who discovered four of Jupiter's moons in 1610 – was born, William Shakespeare lived in an exciting time of discovery and exploration of our complex relationship with the heavens. Shakespeare's work is packed with lunar imagery and direct references to Earth's satellite. He mentions our Moon over fifty times in *A Midsummer Night's Dream* – more than any of his other plays – and most famously, in *Romeo and Juliet*, his titular heroine says to her love: 'O, swear not by the moon, the inconstant moon, / That monthly changes in her circled orb, / Lest that thy love prove likewise variable.'

So how variable is the Moon? From Earth it is both ever-changing and entirely predictable. The Moon appears as different forms in the night sky, depending on its position in relation to the Sun and the Earth. If the sky is clear we see a full Moon about every 29.5 days. This is the length of time it takes for the Moon to go through one lunar phase cycle. It moves from new Moon to full Moon through waxing and waning moons. The time between successive new moons is often called a lunation.

While the Moon's appearance shifts from night to night, its effect on the Earth's seas is regular and precisely foreseeable. Scholars have long speculated about the influence of the Moon on Earth's oceans. The ancient Greeks recognised a relation, and in AD 703 an

English monk, the Venerable Bede, wrote, 'But the most admirable thing of all is the union of the ocean with the orbit of the Moon.' The Moon is the primary control on Earth's tides. Its gravitational pull actually affects the entire Earth, not just the oceans, but because water is much less dense than rocks, this force is most clearly expressed through the rise and fall of the tide. The Moon's gravitational force creates two huge bulges on Earth – one on the side facing the Moon and another on the side facing away from it. As Earth spins on its axis, locations on its surface move through these ocean bulges, resulting in two high tides and two low tides in a twenty-four-hour period. On the opposite side of the Earth from the Moon – where the Moon's gravity is weakest – a high tide is seen because the rest of the solid Earth is being pulled towards the Moon. The difference between high and low tides can be considerable. While the average tidal range worldwide is about a metre, that of the Bay of Fundy on the Atlantic coast of Canada is some 16 metres – the highest in the world. We can see these tidal rhythms imprinted in rocks throughout the geological record.

I experienced a taste of the Moon's power when I was fifteen and part of a school trip cut off by the tide on a beach near Whitby on the Yorkshire coast of England. The water was just above my waist when we were picked up by lifeboat. I suppose we could blame the Man in the Moon as much as my geography teacher, but very precise – and exceedingly useful – information on the times of high and low tides can be found up and down the coast.

What of water on the Moon itself? NASA has invested heavily in the quest to find lunar water because transporting water across space is hugely expensive and inefficient, and finding a source is essential for future human exploration and long-term lunar colonisation. Many thought the Moon was necessarily devoid of water because during the chaos of the planetary impact that created the Moon, volatile substances such as water would have been lost into space. Since the Moon has no atmosphere and its upper surface gets extremely hot in direct sunlight, the only way that water in the form of ice might exist would be in permanently shadowed areas. These are difficult to survey remotely, but data from the NASA

*Lunar Prospector* mission indicated that in fact there may be areas of ice trapped in some craters at the lunar north and south poles. The source of this water is unclear. Many of the impactors that struck the Moon would contain water ice, and if debris reached the deepest craters where temperatures are permanently far below freezing, water ice could potentially survive for 1 or 2 billion years.

The Apollo missions generated many questions about the evolution of the Moon, its internal structure and its relationship with Earth. As I write, plans to return to the Moon with human explorers are well advanced. Finding a local source of water will be a key priority. The Artemis missions led by NASA will be the first crewed lunar landings since *Apollo 17* in 1972. These missions will pave the way to establishing a long-term human presence on the Moon and to sending astronauts to Mars and beyond – the first time humankind will have ventured further than Earth's nearest neighbour.

# CHAPTER 5

# How Old Is the Earth?

The distant past can be difficult for us to comprehend because we experience life in the present. Before the use of scientific dating methods in the middle decades of the twentieth century, there was much controversy and confusion about how old things were. The Greek philosopher Aristotle (384–322 BC) believed that the Earth had always existed. This was a radical view since most ancient civilisations have some kind of origin story. It did not sit well with later scholars who put their faith in literal interpretations of the Old Testament to determine the amount of time elapsed since the Creation.

In the eighth century, the Venerable Bede began the practice of dividing history into Before Christ (BC) and Anno Domini (in the year of our Lord; AD) and came up with a new calculation for the age of the world. He worked out that Christ had been born 3,952 years after the Creation. Bede was accused of heresy by a group of drunken monks, because theologians at that time accepted a gap of 5,000 years from Creation to the birth of Christ. For many centuries, establishment views were so pervasive that some of the most brilliant scientific minds, including Johannes Kepler and Isaac Newton, cited the books of the Old Testament to compute the age of the Earth. The most influential of the biblical chronologies emerged in the middle of the seventeenth century, published in Latin by an Irish priest, James Ussher (1581–1656).

Ussher was a brilliant scholar who became professor of theological controversies at Trinity College Dublin in 1607 and later Archbishop of Armagh and Primate of Ireland. He had become fascinated by calendars and time as an undergraduate and devoted years of scholarship to scouring the Bible and other ancient texts for chronological clues. His *Annales Veteris Testamenti* (*Annals of the Old Testament*) was published in 1650 and began with the declaration that the Creation took place 4,004 years before the birth of Christ. He even worked out the month and the day.

Ussher came to the neat conclusion that the interval between the Creation and End Times (as predicted in the Book of Revelation) was 6,000 years. He then calculated that the first full day of Creation was 23 October 4004 BC. This chronology was hugely influential – Ussher's date for the creation of the world can still be found in editions of the King James Bible. While geological events are rarely dated with such precision, Ussher's chronology dominated thinking about the age of the Earth for the next two hundred years.

Some scholars found the biblical timeframe unsatisfactory and increasingly at odds with scientific observations. Edmond Halley (1656–1741) was a remarkably gifted thinker who is most famous for calculating the periodicity of the comet that bears his name. Halley was born the year Ussher died, and proposed a novel way to estimate the age of the Earth. He argued that Earth's oceans would have been fresh water when they formed, becoming saltier over time as rivers transported dissolved salts weathered from the continents to the sea. He inferred that the salinity of the oceans could be used to estimate their age. If the biblical chronology was broadly correct, the oceans ought to be mostly fresh water. Although Halley did not actually use his method to determine a numerical age for the Earth, he reasoned that the salty oceans were ample proof that the Earth must be *much* older than a few thousand years.

The French naturalist Comte de Buffon (1707–1788) was a radical thinker who carried out novel experiments with cooling iron spheres to model how a molten Earth might cool over time. He estimated the Earth's age to be about 75,000 years. He was a key figure in the eighteenth century as scholars began to contemplate an ancient Earth and deep time.

In the second half of the eighteenth century, James Hutton was the first scholar to show that Ussher's biblical chronology was far too brief to accommodate the vast time depth needed to account for the geological record. Hutton was a keen observer of the natural world, and he studied geological exposures at many locations. One of them is still a key site of pilgrimage for geologists. Siccar Point is a rocky headland poking out into the North Sea about 50 kilometres east of Edinburgh in southeast Scotland. In the summer of 1788 Hutton and two companions visited Siccar Point by boat to examine the rocks exposed in the sea cliffs.

The lower part of the cliff at Siccar Point is formed by grey sedimentary beds that were laid down in an ancient ocean during the Silurian Period, some 435 million years ago. These rocks were later uplifted and folded by tectonic processes so that once-horizontal strata became squeezed to near vertical. These folded rocks were then eroded. The tops of the folds were gradually stripped away before they were capped by red sandstones that formed in a seasonally dry tropical landscape during the following Devonian Period (419.6 to 358.9 million years ago). The dull grey marine rocks and the bright red continental sandstones are thus separated by an unconformity – a term geologists use to describe a gap in the geological record. This unconformity is clear for all to see in the cliffs at Siccar Point. Recent wave erosion has removed some beds of red sandstone, so you can walk on the ancient, Silurian, eroded land surface.

Hutton decoded the sequence at Siccar Point and realised that a great span of time was needed to explain the depositional and erosional features that he saw there. He didn't have knowledge of the true ages – and it would be well into the next century before the Silurian and Devonian periods were even defined – but he explained the sequence of events written in the cliff face during that field excursion in 1788. One of his companions was John Playfair (1748–1819), professor of natural history at the University of Edinburgh. Playfair later wrote down his impressions from Hutton's demonstration: 'The mind seemed to grow giddy by looking so far back into the abyss of time.'

We now know from modern dating that the iconic Siccar Point unconformity represents a break in deposition of about 65 million years. This isn't actually a very big gap in the geological record. At the base of the Grand Canyon there is a gap known as the Great Unconformity that represents over 700 million years of missing information. That stratigraphic gap was first recognised (although not precisely measured) by the geologist John Wesley Powell (1834–1902) during his heroic 1869 expedition along the Colorado river.

Siccar Point has been called the birthplace of modern geology because it provided the inspiration for Hutton and others to see into what became known as *deep time*. While his observations did not allow him to estimate the age of the Earth or even the length of time represented in the strata exposed in the cliff face, Hutton was the first to see that the formation of the geological record demanded vast spans of time that were far longer than any biblical chronology. Hutton also realised that the rock record was punctuated by big gaps. Depending on where one looked, key chapters of Earth history might be missing because the rocks had been eroded away.

Much later, in the 1890s, Irish geologist John Joly (1857–1933) used Halley's salinity theory to generate an age for the Earth. He produced an age of between 80 and 100 million years. Halley's method combined theory and field measurements in a novel way, but it proved too simplistic – it produced an age that was much too young because it did not account for salt losses from the oceans by natural processes. It did show, however, that even rudimentary scientific methods pointed to a very ancient Earth.

The age of the Earth was a hot topic in the last few decades of the nineteenth century and famously brought the disciplines of geology and physics into conflict. Lord Kelvin (William Thomson, 1824–1907) was a hugely respected figure in scientific circles and perhaps the most influential physicist of the nineteenth century. In 1864 he estimated that the age of the Earth lay somewhere between 20 and 400 million years. Kelvin reasoned that, since our planet began as a hot molten mass and had cooled throughout its history, its age could be calculated from measurements of heat loss at the surface. He later narrowed his age calculation to between 20 and 40 million years, but this was far too young for the geologists of the day. How could the immense successions of rock strata and fossils be squeezed even into this chronology? Kelvin's age for the Earth pitted him against the great naturalist and biologist Charles Darwin (1809–1882) because it did not allow enough time for the formation of the fossil record nor for Darwin's grand model of evolution.

At the time, Kelvin was unaware of the large amount of heat that is generated within the Earth by radioactive decay. With our understanding of Earth's internal structure (Chapter 8) and plate tectonics (Chapter 9), we now know that his models for the movement of heat from the deep interior to the surface are unrealistic.

It was the discovery of the radioactive decay of elements that provided the biggest breakthrough in dating the Earth. After decades of passionate debate, physics and chemistry presented geology with a suite of radiometric clocks that could measure deep time.

Most elements in the periodic table are stable – they don't change through time. But some have unstable isotopes that decay through time in a predictable way according to their half-life. By measuring the concentration of these 'parent' isotopes, and the concentration of the resulting 'daughter' isotopes, it is possible to calculate the age of a rock sample using very precise analytical equipment. Several elements and their isotopes have proved useful for dating, depending on the composition of the rock sample and the timeframe of interest.

The rocks that formed Earth's first crust have been completely eroded away or have been recycled into the deep Earth by plate tectonics. This means that even the oldest rocks on Earth's surface are younger than the planet itself. So how can we source samples to inform us about the age of the Earth? To do so we need to examine rocks that formed when the rocky planets of our Solar System accreted. The planets, moons and asteroids in our Solar System are approximately the same age, so we can study lunar rocks and meteorites to help us work out the age of the Earth.

American geochemist Clair Cameron Patterson (1922–1995) was the first to work out the real age of the Earth using meteorite samples. In the 1950s he applied a method called lead–lead dating to establish the age of the famous Canyon Diablo meteorite that smashed into the Arizona desert some 50,000 years ago in the middle of the last glacial period. Patterson published his findings in 1956 in a paper entitled 'Age of Meteorites and the Earth'. The published date was 4.550 billion with an uncertainty of plus or minus (±) 70 million years. This was much older than many in the scientific community had expected or predicted. All radiometric dates involve some uncertainty, but Patterson's analytical work was so rigorous that this date has stood the test of time, and has only been slightly revised. Rock samples collected in 1971 during NASA's *Apollo 14* mission to the Moon have yielded ages of 4.51 billion years.

By establishing the age of the Canyon Diablo meteorite, Patterson had established the age of Earth's accretion. Many more meteorite samples have now been dated, producing ages ranging from 4.53 to 4.58 billion years ago. This is our best estimate for the time when the Earth and other planets in our Solar System formed.

# CHAPTER 6

# Building Blocks: Rocks and Minerals

In my last year at school I devised a project on gravestone weathering. Gravestones are made of different kinds of stone, and they are dated, so I could examine the texture and mineralogy of the various rocks in two local churchyards and measure the extent to which they had disintegrated having been exposed to the elements over a known span of time. Some gravestones had stood for over 250 years. That is a blink of an eye in the great sweep of Earth history, but it got me thinking about how the rocks had formed and the rates at which geological processes might operate. I remember looking at gravestones of granite, sandstone and marble. These fall neatly into the three types of rock that make up the Earth's crust: igneous, sedimentary and metamorphic.

Rocks are made up of minerals such as quartz, feldspar and mica, and minerals are made up of chemical elements such as silica, iron and potassium. There is almost an infinite variety of rock types, but understanding their origin and age is fundamental to our Earth history. The study of rock composition is called petrology, from the Ancient Greek *pétros*, meaning 'rock'. We can study the mineral composition and structure of a rock with the naked eye or by viewing an ultra-thin slice of rock – thinner than a human hair – under a light microscope.

Igneous rocks form from molten rock that has cooled and solidified. Molten rock below the Earth's surface is called magma; when it erupts onto the surface it is called lava. Volcanoes are fed by a magma chamber below ground which supplies the lava that flows out at the Earth's surface. Igneous comes from the Latin *igneus*, meaning 'fiery'. These are hot rocks. Igneous rocks formed the first crust that solidified during the early history of Earth. We will look at the oldest known rocks in Chapter 7.

There are two main types of igneous rock. Rocks that solidify from lava above ground are called *extrusive* igneous rocks. These tend to cool quickly before the crystals in the lava have had time to grow large so tend to have a fine-grained texture; you often need a microscope to see the individual minerals. Rocks that cool very rapidly, such as obsidian, can have a glass-like texture. Basalt is a dark extrusive igneous rock, typically black to dark grey, very fine grained and composed of the minerals pyroxene, plagioclase feldspar and sometimes olivine, depending on the chemistry of the magma. It forms in a variety of volcanic settings – we looked at the extensive basalt lava landscapes on the Moon in Chapter 4. Basalt is the most common igneous rock on the ocean floor but is also common in volcanic landscapes on land. It dominates the landscape of Iceland where the North American and Eurasian tectonic plates are pulling apart, allowing magma to rise to the surface and create new crust. The vast Columbia river plateau covering parts of Washington, Oregon and Idaho in the Pacific Northwest of the USA is a stack of ancient basalt lava flows.

Magmas that solidify below ground are called *intrusive* igneous rocks. These rocks have solidified very slowly in magma chambers located tens of kilometres below the Earth's surface. At such depths, magma is very slow to cool because it is insulated by the rocks around it. Granite is such an intrusive igneous rock, with a coarse texture and minerals visible to the naked eye because they have been able to grow during slow solidification. Granites have a characteristic mix of dark and light minerals in a speckled, grainy pattern. The lighter minerals are quartz and feldspars; the darker minerals are commonly biotite and hornblende. Granites with a pink or reddish hue commonly contain feldspar that is rich in potassium.

We see granites exposed at the surface when the overlying rocks have been eroded away, such as at the famous granite tors of Dartmoor in southwest England. The large interlocking crystals formed during the slow cooling of granite magmas makes them rigid and resistant and ideal for building stone. The massive drums of Nelson's Column in Trafalgar Square in London are Dartmoor granite shipped to London by steamer in 1840. These granites formed around 300 million years ago during a mountain-building phase associated with the formation of the supercontinent Pangaea (Chapter 20) towards the end of the Carboniferous Period. The granite gravestones in my school project did not show any signs of weathering.

Rock classification is full of classical references. A large body of igneous rock that has formed underground is called a *pluton* after the Greek god of the underworld. Plutonic rocks are the most common rocks in the Earth's crust. They form the building blocks of continents and mountain ranges. The spectacular landscapes of Yosemite Valley in California are cut into the hard plutonic rocks of the Sierra Nevada Mountains. The vertical face of El Capitan – a world-renowned challenge for climbers – is a coarse-grained granite that solidified in a huge magma chamber during the Cretaceous Period. The colossal heads of four United States presidents – George Washington, Thomas Jefferson, Abraham Lincoln and Theodore Roosevelt – at the Mount Rushmore National Memorial in South Dakota are carved in granite that formed some 1.8 billion years ago during the Proterozoic Eon.

Sedimentary rocks, meanwhile, are formed from material that has eroded from pre-existing rocks or from the accumulation of once-living organisms. They are especially important in Earth history because they contain most of the fossil record. Finding sedimentary rocks from the early part of the Archean Eon (4 to 2.5 billion years ago) is also exciting and important because they can tell us if oceans were present and whether sediments were deposited by rivers. From this we can glean information about the nature of the atmosphere and when a fully functioning global water cycle with landmasses, lakes and rivers had become established.

The names of sedimentary rocks commonly reflect the dominant size of the particles they contain, such as sandstones and mudstones. The latter rocks often form in low-energy environments such as the deep ocean floor or the bottom of a deep lake. Such settings can provide good conditions for the preservation of plant and animal fossils, as we shall see in later chapters. Indeed, chalk is a sedimentary rock that is completely dominated by the remains of microscopic marine organisms that built up on ancient seafloors.

Contrary to their solid, timeless appearance and popular associations, the rocks of the Earth's crust are in a continuous cycle of transformation and renewal. The rock cycle is the ongoing process of rock formation and rock breakdown driven by a range of mechanisms. For instance, mountains are eroded by glaciers and rivers and other processes, producing sediments that are ultimately transported to the ocean where they are deposited to eventually form new rocks. When these rocks are deeply buried they are transformed by heat and pressure. The rock cycle continually reshapes the Earth's surface over millions of years, creating new rocks and breaking down existing ones.

Sedimentary rocks are made up of materials that have been transported by water, wind, ice or gravity. The presence of sedimentary rocks is evidence of an active rock cycle with material eroded from one location and then transported and deposited in another. As the sediments and fossils build up they are squeezed together and compacted to form hard rock – in other words, they become *lithified* (from the Ancient Greek *lithos*, meaning 'rock'). In sandstones, the grains are often cemented together by a material that is different from the main particles, making it a generally softer and naturally porous rock. I saw the effects of this in my school project.

The red sandstone gravestones were the most heavily weathered – the inscriptions were fading, and grains of red sand had fallen to the base of the gravestone. In this case the carbonate cement was the weakness; while the quartz sand grains were very resistant, the cement was susceptible to chemical weathering by acid rain.

Metamorphic rocks are the final category. These are rocks that have been transformed by the action of heat or pressure or both. As rocks are buried they may be squeezed and folded by tectonic processes; they may also come into contact with magma. There are two main types of metamorphic rocks. Some, known as *foliated* metamorphic rocks, have a distinctive banded or layered fabric – slate, schist and gneiss are good examples. Some of the oldest rocks on Earth are metamorphic rocks. These rocks have sometimes been so intensely modified – both physically and chemically – that it is not possible to identify the original parent rock, or *protolith*.

Marble is an example of a non-foliated metamorphic rock. In this case we know what its protolith commonly is: limestone. Similar to chalk, its softer and more finely grained cousin, limestone is a sedimentary rock mainly composed of the mineral calcite ($CaCO_2$), often derived from the accumulation of tiny shells or corals on the ocean floor. During deep burial and heating the calcite recrystallises and the texture of the rock is transformed so that the original fossils are no longer recognisable, and are instead replaced by growing and increasingly interlocking crystals of calcite. The limestone has become marble.

Fine-grained marble can be sculpted and polished. The Carrara marble quarries in the hills of northern Tuscany have been a highly prized source of marble since pre-Roman times. And the marble itself has a long history: the original limestones were laid down in a warm tropical ocean during the Cretaceous Period and were metamorphosed into marble during the tectonic deformation associated with the uplift of the Alps and Apennine Mountains during the Oligocene and Miocene epochs. The Romans used Carrara marble to decorate the Pantheon and to form the drums of Trajan's triumphal column in Rome, and Michelangelo and Bernini sourced their marble in the Carrara quarries during the Renaissance.

Closer to home, I looked at marble gravestones in my school project. They showed signs of weathering by solution, because rainwater is a mild acid that attacks calcite. This is also the reason why those with marble kitchen worktops wince if vinegar or lemon juice is spilt on them.

Slate is a very fine-grained metamorphic rock. It has thin foliation along which it readily splits to leave smooth, flat and resistant surfaces. The famous slates of Snowdonia (Eryri) were originally mudstones deposited in a deep ocean in the Cambrian Period some 500 million years ago. The mudstones were uplifted, squeezed and folded in the Ordovician to form slates. The slates on the roof of my house in south Manchester came from North Wales in the 1880s. These Welsh quarries 'roofed the world' in the nineteenth century; their slates were exported to almost every continent.

Igneous, sedimentary and metamorphic rocks are chronicles of Earth history. But the way that the rock record should be interpreted was a matter of great debate in the eighteenth century. The highly influential German mineralogist Abraham Gottlob Werner (1749–1817) argued that all rocks and minerals found in the Earth's crust had formed in a universal ocean by chemical precipitation and then by the deposition of sedimentary rocks. He argued that granite, for example, with its crystalline structure, had precipitated from seawater. This universal ocean eventually receded to leave the continents high and dry. His philosophy become known as *Neptunism* after the Roman god of the sea. Werner's ideas were remarkably influential at the time, and he had many followers across Europe.

An opposing group, who became known as *Plutonists*, stressed the role of igneous activity and Earth's internal heat in uplift and mountain formation. They also argued for many cycles of gradual erosion of igneous rocks to form sedimentary layers. Scottish geologist James Hutton (whom we met in Chapters 1 and 5) was most closely associated with these opposing ideas, which eventually won the day.

The concept of the rock cycle was first put forward by Hutton at the end of the eighteenth century. He was a keen observer of the natural world and made seminal contributions to establish geology as a rational science. As he developed his thinking around the continuous cycling of sediments from the uplands to the sea, Hutton famously stated, 'We find no vestige of a beginning, no prospect of an end.'

Hutton was the first geologist to make convincing arguments about the reality of deep time in Earth history. He was also a champion of what became known as *uniformitarianism*, the view that geological processes that could be observed in the present had operated in the same way in the past. Change happened slowly by the cumulative work of observable processes. Hutton rejected the idea that catastrophes such as the biblical flood played a role in accounting for the rock record. Instead he showed that Earth was an ancient planet and argued forcefully that the interpretation of ancient rocks should be based on the study of modern-day processes. His ideas were developed and championed most notably by Charles Lyell (1797–1875) in the nineteenth century (Chapter 24).

One of the key tasks of the geologist attempting to work out the geological and environmental history of an area is to map the extent and thickness of the rocks and establish the relationships between them. This is known as stratigraphy. At the dawn of the industrial age it was key to establishing the location and extent of resources such as coal, iron ore and building stones.

William Smith (1769–1839) was a pioneer of stratigraphy and geological mapping and the first person to systematically map the geology of an entire country. 'Strata Smith', as he became known, was employed as a surveyor for the growing canal network in England. He worked out that layers of rock in different areas could be matched up using information on their properties and the fossils they contained. In 1815 he published a beautiful geological map of England and Wales with different colours representing the various strata, many of them the same hues as the rocks themselves. Even though Smith's work was fundamental in establishing the principles of geological mapping and working out the nature of the geological record, it took many years before his discoveries were properly recognised by the geological establishment. As the son of a blacksmith, he did not fit easily into scientific circles, and he had to overcome many difficulties during his life. His map was often plagiarised, and bankruptcy landed him in prison for ten weeks in 1819. Sixteen years after the publication of his famous map, the Geological Society of London awarded him its highest honour.

Rock appreciation is part of human history. We have always looked at rocks and carefully selected them for their useful properties. They are also key to unravelling Earth's story. To understand the origin of our planet and the evolution of life, we must read the rocks.

# CHAPTER 7

# The Oldest Rocks on Earth

Rocks and minerals are time capsules. They store information about the environmental conditions that prevailed during their formation, and so provide the record that tells us the story of Earth's development through time.

To glimpse the origins and very early development of the surface of our world we need to examine the remnants of Earth's primordial crust. But therein lies a problem – almost all of the early geological record has been removed. Little hard evidence survives from the first half-billion years of Earth history because these rocks have been destroyed during more than 4 billion years of mountain building, erosion and tectonic activity. The name of the Greek god Hades is often taken to mean 'the unseen one' – so too, the vast majority of the early crust from the eon that bears his name is long gone.

Despite their rarity, however, remnants of early Earth's crust *have* been discovered by geologists. These rocks are difficult to study because they have been subjected to many episodes of intense deformation, folding and metamorphism. Earth's most ancient terrestrial materials – found most notably in Canada, Greenland, Antarctica and Australia – are an exciting research frontier and have generated enormous interest and much controversy. What can they tell us about the early Earth?

As we have seen, the Hadean Eon (4.567 to 4.031 billion years ago) witnessed the initial stages in the formation of the Earth when huge volumes of dust and gas circling the young Sun were pulled together by gravitational forces to form increasingly large fragments, and eventually our planet. This accretion phase saw the formation of a dense core as the heaviest molten elements sank into the interior with the lighter elements forming a mantle and, eventually, the first rocky crust. The Earth was continually bombarded by meteorites during this period, adding mass to the young planet. These impacts were so intense that the heat released kept much of the Hadean Earth's surface in a molten state, covered by oceans of basalt.

Early Hadean times were, indeed, hell on Earth. Any solid rocks that crystallised may have been rapidly consumed into this magma ocean. As the surface temperature of the Earth cooled, the earliest surviving crust will have been composed of igneous rocks that solidified from magma. We have lots of theories about the nature of Hadean Earth but very few rock outcrops to guide our thinking. Have any rocks survived from the Hadean? This question has been hotly debated, but scientific dating is helping geologists to piece together what remains of that Hadean world.

It is not easy to comprehend the antiquity of these most ancient rocks. The Hadean is followed by the Archean and Proterozoic Eons; together these three vast spans of time make up the Precambrian, accounting for over 4 billion years (about 88 per cent) of Earth history.

The Hadean was originally proposed in 1972 as an informal name to cover the period of geological time that *preceded* the formation of the earliest known rocks, and its upper boundary was set at 4.0 billion years ago. But as our understanding of Earth's ancient rocks has progressed and scientific dating has been refined, a handful of sites from far-flung locations have yielded ages close to and even older than 4.0 billion years ago.

The only way to establish the age of these very old rocks is by making precise measurements of the radioactive isotopes trapped inside their constituent minerals. Because radioactive isotopes decay at a known rate according to their half-life, they can provide a geological clock. The results are often called radiometric ages. In these very old rocks, the most common measurement used to determine a mineral's age is the ratio between radioactive uranium and a stable daughter lead isotope (from the end of the radioactive decay chain). But to do this, you need a very stable mineral – one that can keep its isotopic signature for 4 billion years. Luckily, zircon is just such a mineral, and it is fairly common in the Earth's crust, found in igneous, metamorphic and sedimentary rocks.

Zircon is a hard, dense and chemically stable mineral at the Earth's surface, and remarkably resistant to weathering. Zircon crystals can be formed and then eroded, transported, deposited, incorporated into new rocks and reworked multiple times with little or no alteration of their isotopic values. This remarkable stability means that zircon crystals can retain information on their age even when the original rocks they formed in (often granitic or metamorphic rocks) have long disappeared. Mineral grains that have been reworked and incorporated into younger sedimentary rocks are called *detrital* grains. And detrital zircons are special because they can retain information on their original age of formation as well as the environmental conditions that prevailed when they were formed. These tiny crystals provide a direct link to what our early world was like.

While crustal rocks that give Archean ages are surprisingly extensive and present on all seven continents, only a handful of these areas have been shown to contain rocks older than 3.7 billion years. These masses of very old rocks are often called shields or cratons, from the Greek *kratos*, meaning 'strength'. As more radiometric dates become available, blocks of even older crust are still being found within these cratons. It is highly likely that blocks of very old crust lie beneath the great ice sheets of Greenland and Antarctica. Indeed, Soviet geologists working in East Antarctica in the 1970s reported ages of 4 billion years for rocks in the Fyfe Hills of Enderby Land near Casey Bay. Although these results have generated some controversy, more recent work on the highly deformed gneisses of Enderby Land has revealed ages of over 3.8 billion years.

The rocks making up the ancient cratons are often so strongly deformed that it can be extremely difficult to unravel the original relationships of the various rock units. Some are *metasedimentary* rocks, which means they are sedimentary rocks that have been subjected to elevated temperatures and pressures so that the original grains have recrystallised, but the key sedimentary structures of the original rock can still be recognised. The presence of such ancient sedimentary rocks is significant because they can inform us about early Earth surface processes and climate, especially if flowing water was involved in their formation.

At the time of writing, the oldest crustal rocks on Earth that have been isotopically dated come from the Acasta Gneiss Complex in northern Canada, just south of the Arctic Circle. These rocks were discovered in the early 1980s during regional mapping by the Geological Survey of Canada, and form part of the Canadian Shield, the largest expanse of Precambrian rocks on Earth. In the tundra wilderness of the Northwest Territories, the exposed assemblage of strongly deformed and metamorphosed rocks forms beautifully contorted banded outcrops along the Acasta river. Rocks in the Acasta gneisses range in age from 4.03 to 3.6 billion years old, and they were originally granites within Earth's early crust.

On the eastern side of Hudson Bay, some 35 kilometres south of the Inuit village community of Inukjuak, metamorphosed volcanic and sedimentary rocks make up what is called the Nuvvuagittuq Greenstone Belt; it is the oldest part of the Canadian Shield. Geologists from the University of Ottawa have mapped and dated these rocks. Metamorphosed rocks from this site have yielded ages of 3.82 to 3.75 billion years (the early part of the Archean Eon), with the precursor rocks of these greenstones forming some 4.3 billion years ago. Some features of the minerals in this primitive crust also indicate they were altered by super-heated seawater, so a water cycle must have been in place. These rocks also provide good evidence for the formation of oceanic crust during the Hadean (in an early ocean) and perhaps some early plate tectonic activity within 200–300 million years after the formation of the Earth – much earlier than previously has been thought. There is still a lively debate about when plate tectonic processes began and the style and extent of tectonic activity that might have emerged in the Hadean Eon (Chapter 9). Evidence for the presence of deep water implies that not all of the Hadean was hell-like.

A well-preserved block of early Archean crust is also found across the Labrador Strait in the Nuuk region of southwest Greenland. The Itsaq Gneiss Complex has yielded ages of 3.9 to 3.6 billion years, close to the Hadean/Archean boundary. It is the best-described record of early crustal evolution that we have, and suggests that the oldest preserved continental crust was created in a setting where tectonic plates converged. The Isua Greenstone Belt in southwest Greenland is an especially important part of the early geological record because these rocks may contain some of the earliest known evidence of life at 3.7 billion years ago in the form of layered mats of microbes called stromatolites (Chapter 11) – although not everyone agrees with this interpretation. There are features that resemble stromatolites, but they may be the result of intense tectonic deformation rather than biology.

Dating the oldest rocks on Earth is not always straightforward. The isotopic measurements used in dating come with analytical errors and this uncertainty still leads geologists to interpret the results in different ways and to argue about which are the world's oldest rocks.

As we have seen, the remarkable stability of zircon minerals means they retain their original age of formation. Looking at reworked zircon grains contained in sediments in these very early crustal rocks reveals even older ages for the original formation of the mineral. The oldest and largest number of detrital zircons come from metamorphosed conglomerates in the Jack Hills region of Western Australia. A conglomerate is a sedimentary rock containing rounded pebbles within a fine sediment matrix and usually hardened like concrete. The Jack Hills conglomerates are about 3 billion years old, from the Archean Eon, and because they are rich in zircon crystals, they hold deeper memories that reach back into the Hadean – they are the richest source of Hadean material currently known on Earth. These zircon crystals have been dated using uranium–lead isotopes to yield an age of 4.36 billion years. They are the oldest known terrestrial material, and hard evidence for the existence of *continental* crust during the Hadean.

The very hard, resistant and predominantly crystalline rocks that sit in the lower part of the Earth's crust, and which have largely been the focus of this chapter, are conventionally called *basement* rocks. Petroleum geologists often mapped these as 'the crust below rocks of interest', or labelled them as 'non-prospective' rocks, because they did not hold oil or gas. But how times have changed! Many basement rocks are now of enormous interest since they can be rich in rare earth elements needed for the green energy transition.

The Greenbushes open pit mine in Western Australia is the largest hard rock lithium mine in the world. The very coarse-grained Archean granites (known as pegmatites) of this region are enriched with the lithium-bearing mineral spodumene. Lithium is a key raw material for the manufacture of rechargeable batteries and thus of immense importance for the decarbonisation of energy and transport to meet global net zero ambitions (Chapter 37). The Greenbushes operation is expanding to meet deficits in the global lithium supply chain. Earth's ancient crust is now big business.

Earth is some 4.56 billion years old, and the oldest minerals dated by geologists are about 4.3 billion years old. Although rare, the discovery of Hadean rock samples provides a tantalising view of what may have been Earth's first terra firma. Since their properties indicate that tectonic plates may have been in existence since the Hadean, it is remarkable that any of these rocks have survived the ravages of geological time. Working out the geological history of the early Earth means unravelling the very complex signatures in ancient rocks subjected to multiple phases of intense metamorphism and deformation. There are formidable challenges in studying the oldest rocks in the Earth's crust, but it is of fundamental importance for piecing together the opening chapters in Earth history, including evidence for the earliest life.

# CHAPTER 8

# Journey to the Centre of the Earth

It has often been said that we know more about the surface of the Moon than we do about the materials beneath the Earth's crust. Three hundred years ago Isaac Newton calculated that the average density of planet Earth is approximately twice that of its surface rocks, demonstrating that the interior must contain much denser material. Newton's reasoning was sound, but it would be another two centuries before geologists had the tools to work out the nature of these materials and how they were arranged. In 1864 Jules Verne published *Journey to the Centre of the Earth*, the epic tale of Professor Lidenbrock and his companions, who explore a fantastical and perilous underworld of dinosaurs and subterranean oceans. Verne's book fired the public imagination but also highlighted how little was known about Earth's deep interior. What lies at the core of our planet? Could life exist deep below the Earth's surface? How far could humans travel into this underworld?

After its formation some 4.56 billion years ago, Earth began a long and complex process of matter differentiation, in which the heaviest elements sank deepest into the Earth, creating concentric layers of crust, mantle and core. We cannot send probes into the mantle or core because of extreme temperatures and pressures – the core–mantle boundary is as hot as the surface of the Sun. Scientists can only detect these layers indirectly, using seismology – the study of seismic waves generated by earthquakes (and large explosions) and the properties of the materials through which they travel. It uses geophysical data to track earthquake activity and build images of the Earth's interior. Shortly after Verne's book was translated into English in the 1870s, the first modern seismometers – which can now detect the arrival of seismic waves from across the globe – were invented, and the scientific study of Earth's interior began to take shape.

In April 2022 a bronze statue was unveiled in a park in Zagreb in tribute to the achievements of one of Croatia's greatest scientists. Early in the twentieth century, Andrija Mohorovičić (1857–1936) discovered the boundary between the Earth's crust and mantle, named for him the Mohorovičić discontinuity, or simply Moho. Mohorovičić is often called the founder of modern seismology; his statue sits in front of the building where he installed a seismograph in 1906.

In 1908 Mohorovičić acquired new equipment that made the earthquake observatory in Zagreb one of the most advanced in Europe. The following year an earthquake struck the Kupa Valley about 35 kilometres southwest of the city. Mohorovičić located the epicentre and wrote to colleagues across Europe's rapidly expanding seismological network to find out how the earthquake had been recorded at their locations. The most distant seismic record came from Tbilisi in Georgia some 2,400 kilometres away. In little over a year, Mohorovičić had analysed all the seismic records, formulated a new model of seismic wave propagation, and published his findings in both Croatian and German. Mohorovičić was the first to demonstrate that the velocity of a seismic wave is influenced by the density of the material through which it passes. Seismic waves tend to move faster in denser, more rigid material. They are faster in solid rock than in loose sediment, for example.

To reconcile the physics of wave behaviour with the seismic dataset he had assembled, Mohorovičić proposed the existence of a distinct break in the characteristics of the rocks at about 50 kilometres' depth, where he observed a jump in seismic wave velocity. This jump marked the boundary between crust and mantle. This elegant demonstration was a crucial step in unravelling the structure of our planet.

Earth has a cold and rigid crust of two kinds: continental and oceanic. Below the oceans the crust has an average thickness of about 7 kilometres. It is much thicker on the continents, where it averages about 35 kilometres. Oceanic crust is rarely much older than about 200 million years because it is destroyed at subducting plate boundaries where it sinks into the mantle. In contrast, the oldest continental crust is very ancient indeed; it can be more than 4.0 billion years old – almost as old as the Earth itself. The thickness of the continental crust mirrors the elevation of its landscapes, ranging from about 10 to 75 kilometres. It is thinnest beneath low-lying plains and thickest below high mountains. On the Indian subcontinent, for example, the average crustal thickness across the great lowland floodplains of the Ganges and Brahmaputra rivers is 30–35 kilometres, rising to 60–80 kilometres in the Himalayas. On average, the temperature in the crust increases with depth by some 30°C per kilometre.

The mantle is by far the thickest layer, with its base lying about 2,890 kilometres below the Earth's surface. It makes up 84 per cent of Earth's volume. The mantle is mostly solid rock dominated (97 per cent) by a mineral group known as silicates, which contain the elements silicon and oxygen. Silicate minerals include olivine, pyroxene and garnet. The most important rock in the upper mantle is called peridotite, a dark green and coarse-textured rock dominated by olivine and pyroxene. Both temperature and pressure increase with depth in the mantle. Temperatures range from about 1,000°C below the Moho to almost 4,000°C at the boundary between lower mantle and super-heated core. Heat is transferred from the core to the upper mantle by the process of convection. The base of the mantle is heated by the core, so material begins to rise. The cooler material at the top of the mantle sinks to create a series of giant convection loops. Imagine a giant lava lamp with huge blobs of ductile rock being deformed in super-slow motion. This process takes place on a such a grand scale it is powerful enough to move Earth's tectonic plates (Chapter 9).

Richard Oldham (1858–1936) came from a family of geologists and geographers and led the Geological Survey of India. At about the same time that Mohorovičić was developing an interest in seismology, Oldham studied seismic data from the Great Indian Earthquake of 1897 and noticed that the European records showed three distinct wave forms. He also observed that seismic waves travelled at different velocities through the centre of the Earth, leading him to conclude that it had a quite different composition to the mantle and crust. Oldham had discovered Earth's core but was unsure about its composition.

Oldham was the first to recognise primary (P) and secondary (S) waves. P-waves are the fastest type of seismic wave produced by earthquakes; they can travel through solids, liquids and gases and are the first waves to arrive at seismic recording stations. P-waves help us to locate earthquakes and understand Earth's internal structure. S-waves can only travel through solids, and they travel more slowly than P-waves, but cause more shaking and damage due to their stronger side-to-side or up-and-down motion.

For a few decades after Richard Oldham's findings it was widely held that the entire core was molten, until Danish seismologist Inge Lehmann (1888–1993) conducted a detailed study of the seismic waves generated by a major earthquake that struck the South Island of New Zealand in June 1929. From their arrival times in Denmark and Greenland, Lehmann deduced that some of the waves must be bouncing off an even deeper surface. She was able to show that Earth's core has two parts: a liquid outer core and a solid inner core. Her discovery of a solid inner core was the last big piece in the puzzle of Earth's deep interior. Lehmann's theory was confirmed in 1970 when more sophisticated equipment recorded waves deflected by a solid core.

The inner core is about the size of the Moon with a radius of some 1,200 kilometres. It is the hottest part of our planet – temperatures can reach 5,500°C – and a solid mass of mainly iron and nickel. Metallic iron is denser than rock, so it sinks towards the centre. It is believed that all of the rocky planets in our Solar System (Mercury, Venus, Earth and Mars) have an iron-rich core. Earth's outer core is a hot liquid mass of iron with some nickel and lighter elements. The liquid core is the source of Earth's magnetic field (Chapter 12). Earth's core is not a perfect sphere and its boundary with the mantle is probably quite irregular.

The age of Earth's core and the speed that iron was able to move to Earth's centre are both disputed. Some Earth scientists argue that the iron sank rapidly soon after the Earth formed; others think it was a slow trickle downwards over an extended period. Since the oldest rocks in the crust preserve evidence of a magnetic field, the core must have begun to form fairly early in the evolution of our planet and at least 3.5 billion years ago. Data from the decay of radioactive elements suggest the inner core began to harden some-time between 1 and 2.5 billion years ago.

Planets are hot at depth because they store enormous amounts of heat left over from accretion. Heat is also generated from the decay of radioactive isotopes and from heat that is released as the liquid outer core slowly solidifies into solid iron. The origin of the core is one of Earth's deepest mysteries, but most researchers agree that the solid inner core formed by crystallisation of a liquid core.

The rigid outer layer of the Earth, which includes both crust and upper mantle, is known as the lithosphere. It is the coolest and most rigid of Earth's layers, extending to a depth of about 100 kilometres. Below is the asthenosphere, a weaker layer (*asthenós*, 'without strength') of the mantle that deforms comparatively easily and above which the lithospheric plates move. The thickness of the astheno-sphere ranges from around fifty to several hundred kilometres; its lower boundary is uncertain. The terms lithosphere and astheno-sphere describe the physical behaviour of the associated rocks but not their composition. The movement of heat and matter in the mantle drives plate tectonic processes in the lithosphere (Chapter 9) and therefore helps shape the arrangement of continents and ocean basins as well as the landscapes of the Earth's surface.

Though we understand much more about the deepest inner workings of our planet, humans will never journey to the mantle, never mind the centre of the Earth. Despite many years of effort and a Cold War race, not even the most advanced equipment has come close to drilling through normal crust to the mantle. In the 1960s a Soviet team began drilling through the crust on the Kola Peninsula inside the Arctic Circle in an attempt to sample the zone where crust and mantle interact. They drilled over 12 kilometres into the crust, further than anyone before or since. It took twenty-four years to reach that depth, and the project was abandoned in 1992. The furthest that humans have vertically ventured into the crust was at the Veryovkina Cave in the Western Caucasus of Georgia, the deepest known cave on Earth. Having narrowly escaped flood-filled passages during a previous descent, Russian speleologists reached its record depth of 2,212 metres in August 2019.

The global network of seismometers expanded rapidly after the Second World War to help monitor the testing of nuclear weapons. This has produced big seismic datasets and allowed the development of incredibly detailed 3D models of Earth's interior. A century on from the discovery of the Mohorovičić discontinuity, satellite sensors can now detect the tiny variations in Earth's gravitational field that result from the rock density contrast between mantle and crust. Unravelling the workings of Earth's deep interior is a critical part of better understanding the evolution of its landscapes and the living systems on the surface.

# CHAPTER 9

# Plate Tectonics and Shifting Continents

In 1835, during his epic journey on the *Beagle*, the recently gradu-
ated Charles Darwin witnessed a violent earthquake in the coastal
town of Concepción in central Chile. The geological fault that
slipped lifted beds of mussels more than 3 metres above the water
line. A few weeks later Darwin observed similar mussel beds
hundreds of metres above sea level in the Andes. These experiences
convinced him that Earth's crust was dynamic and that earthquakes must play a role in the formation of mountain ranges.
While he also realised that this required a huge span of time, he
was unable to explain the underlying forces.

Tectonics is the branch of geology concerned with the structure
of a planet's rigid outer shell and the origin of features such as
faults, folds and mountains. It comes from the Ancient Greek
*tektonikos*, relating to construction, and includes the study of forces
that operate deep in the Earth's interior as well as those that shape
the landscapes at the surface.

Explaining the formation of mountains was a major puzzle for geologists well into the twentieth century. The continents were considered to be static and there was no satisfactory explanation for marine fossils found thousands of metres above sea level. We know now that mountain ranges are formed by the Earth's crust buckling under the enormous stresses produced by the collision of the giant tectonic plates that form Earth's outer shell. They are the work of a dynamic planet.

On 6 January 1912, at a conference held in the Natural History Museum in Frankfurt, a German meteorologist and polar scientist presented a radical new theory. Alfred Wegener (1880–1930) was not the first scholar to notice the jigsaw-puzzle-fit of the continents on either side of the Atlantic Ocean, but he was the first to use the term *continental drift* and to publish a fully developed theory on the formation and breakup of a supercontinent. Wegener put forward the idea that the continents were not static but moved slowly across the Earth's surface. He argued they had once formed a supercontinent that he later called Pangaea.

There were several lines of evidence that suggested the distribution of continents and ocean basins was not fixed. It wasn't just the shape of the continents. Wegener pointed to ancient rock formations and groups of fossils that matched up on either side of the Atlantic Ocean. He published *The Origin of Continents and Oceans* in 1915 but, as the First World War raged across Europe, his book and its bold theory had little impact.

The geological community ridiculed these new ideas. Geologists in the United States were especially hostile after Wegener presented a seminar on his theory in New York in 1926. The German sat quietly listening to the negative comments and smoked his pipe. It did not help that Wegener was not a geologist, but his failure to offer a realistic mechanism that could explain the movement of continents was a major weakness in his hypothesis. Continental drift was out of the question for many geologists. They saw Earth's interior as far too rigid to allow any sideways movement. No force even remotely powerful enough to move the continents across the Earth's surface was evident.

After the Second World War, the tide began to turn. Surveys of the ocean floor funded by the US government revealed a global network of submarine ridges and mountains associated with earthquakes and volcanic activity. This network of mid-ocean ridges stretched for thousands of kilometres. In 1963 two British geologists, Frederick Vine (1939–2024) and Drummond Matthews (1931–1997), published the crucial data that confirmed sea floor spreading. They showed that the ocean floor increased in age in both directions as you moved away from these ridges. They neatly matched up rocks of the same age on either side by measuring alternating shifts in magnetic polarity. As lava emerged and solidified, the oceanic crust moved away from both sides of the ridge where it was forged. The continents really did drift apart.

By the end of the 1960s it had been established that the lithosphere was made up of seven major and eight minor plates – the Pacific, North American and Eurasian plates are the largest – and a series of microplates. The lithospheric plates are gigantic slabs made up of the crust plus the underlying layer of cold, dense brittle mantle that can be up to 250 kilometres thick. They carry both continents and oceans across the surface of our planet.

The tectonic plates are in continual motion, and the margins of the plates are some of the most hazardous places in our world. Some plates are moving together, some moving apart, others sliding and scraping alongside. Plate boundaries are classified into three main types based on these movements.

Where tectonic plates collide (convergent boundaries) the crust is buckled and pushed upwards, and mountain ranges may form. The Indian plate barged into the Eurasian plate some 50 million years ago to build the mighty Himalayas (Chapter 26). The summit of Mount Everest is 8,850 metres above sea level. The hard limestone rocks that form the highest point on Earth contain fossils of Ordovician trilobites and brachiopods, creatures that lived on the deep ocean floor some 470 million years ago.

Oceanic lithosphere is created at mid-ocean ridges where tectonic plates move apart, and basalt lava emerges via sea-floor fissure eruptions. These are called *divergent* or *constructive* plate boundaries. The mid-Atlantic ridge is the boundary between the North American and the Eurasian plates in the North Atlantic, and it separates the African and South American plates in the South Atlantic. It was first identified in the 1950s by Marie Tharp (1920–2006), who used sonar data collected from research vessels to plot the first detailed map of the Atlantic Ocean floor and fundamentally changed our understanding of ocean-floor topography. Although her findings were initially dismissed, she is now recognised as one of the foundational figures in modern Earth science and oceanography whose work helped to confirm the existence of plate tectonics. The mid-Atlantic ridge extends over 16,000 kilometres from the Arctic Ocean to inside the Antarctic Circle; it extends below Africa and becomes the boundary between the African and Antarctic plates in the Indian Ocean. It is a mostly sub-marine mountain range rising between 2,000 and 3,000 metres above the ocean floor. Iceland is a rare place where this mid-ocean ridge sits above sea level. In Thingvellir National Park southwest of Reykjavik you can jump across from the North American to the Eurasian plate.

A *subduction zone* is where a tectonic plate sinks beneath another and the descending slab is eventually assimilated into the mantle. Deep marine trenches form in these settings, with chains of volcanoes on the overlying plate fed by magma from the recycled plate. The deepest parts of the ocean are found in subduction zones. The southern end of the Mariana Trench in the western Pacific is where the western edge of the Pacific plate plunges beneath the Mariana plate. Known as the Challenger Deep, the seabed lies some 11 kilometres below the water surface – the deepest part of the global ocean.

The rocks in a sinking plate take millions of years to reach the same temperature as the mantle but the plate can release huge volumes of super-heated water that forces melting in the rocks above. Such water-rich magma can lead to explosive volcanic eruptions – the bubbles behave like the gas in a fizzy drink. The highly explosive 1991 Mount Pinatubo eruption in the Philippines was one of the largest of the twentieth century, forcing the evacuation of 58,000 people. Pinatubo sits on the subduction zone where the eastern edge of the Eurasian plate slides under the Philippine Sea plate.

The edges of the giant Pacific plate are known as the Ring of Fire because of the dense concentration of tectonic hazards. More than 450 volcanoes and over 90 per cent of Earth's earthquakes are associated with the subduction zones that fringe this plate.

The third type of plate tectonic boundary is known as a *transform fault* where plates grate past each other. These are known as conservative margins because lithosphere is neither created nor destroyed. Transform faults are mainly found on the ocean floor, especially where one length of mid-ocean ridge is offset from another. Transform faults are also present on land – the San Andreas Fault in California is perhaps the best-known example. It forms the boundary between the Pacific and North American plates where the enormous stresses that build up can lead to catastrophic earthquakes.

The upper part of the lithosphere is made up of either oceanic or continental crust. Oceanic crust forms at mid-ocean ridges where basalt lavas emerge. Continental crust is mainly formed along subduction zones where partial melting leads to the formation of rocks rich in silica – such as andesites and granites – in the volcanic provinces on the overriding plate. This process produces crust that is much thicker and less dense than oceanic crust and thus not so easily subducted. Basalt is denser than granite so oceanic lithosphere sinks below continental lithosphere. The continents are very ancient indeed because they are difficult to subduct into the mantle. All the continents contain extensive rocks of Precambrian age, yet the oldest oceanic crust is much younger, about 180 to 200 million years old. Some of the oldest oceanic crust can be found on the western edge of the Pacific plate close to the Mariana Trench and on either side of the central Atlantic Ocean.

We know that the lithospheric plates move just a few centimetres each year, at about the same speed as the growth of your fingernails. But how do they move? Their movement is the surface expression of mega-scale processes taking place in Earth's hot dynamic interior. Heat moves through the mantle by convection from the super-hot core to the cool, hard, brittle crust. Rocks in the mantle are more plastic and more easily deformed than crustal rocks. Despite the extreme temperatures in the mantle, rocks remain solid because of the immense pressures placed upon them. Over extremely long timescales, however, mantle rocks will flow. Like tar warming in a cauldron, with hotter material rising and cooler, denser material sinking, these mantle rocks ooze and flow exceedingly slowly. The temperature contrasts between the lower and upper mantle create giant convection cells with a radius of thousands of kilometres. These immense physical forces slowly churn the mantle rocks and thus drive the horizontal movement of lithospheric plates at the Earth's surface. The lithospheric plates do not float on the mantle; they are part of the mantle convection process that releases heat from the interior.

When did plate tectonic processes begin? Earth scientists have debated this question since the theory gained wide acceptance in the late 1960s. They have searched the early rock record for tell-tale signs of plate movement because it is important to understand how this planetary-scale rock recycling scheme fits into Earth history. There has been much discussion about the style of tectonic activity on early Earth. It may have followed what has been called the stagnant lid model, with a non-moving outer shell that is too rigid to participate in the convection flows operating in the mantle. Under these conditions, processes in the mantle would have still caused volcanic activity but not large, global-scale tectonic plate movements. There would have been far fewer earthquakes too.

There is a surprisingly wide range of views on the inception of plate tectonics. Some argue for a very early onset in the Hadean, about 4.2 billion years ago. Others put the onset at 3 billion years ago with some arguing for a very late onset at a billion years ago. Plate tectonics did not begin overnight. It was almost certainly an evolutionary process with starts and stops – a process that perhaps looked rather different in style and extent during earlier stages of Earth history before it became a planetary-scale feature. When the young Earth began to cool down and the first crust formed, the

Earth would have been a single-plate regime. The first appearance of plate tectonics may have been on a local scale with small plates becoming mobile before breaking up and being consumed into the mantle.

Earth is the only planet in our Solar System with active plate tectonics. We can add this to the list of features that make our world unique. Some planets have a history of volcanic activity, and the Moon has huge expanses of terrain where basalt lava once flowed, but no other planet or moon in the Solar System has ever experienced planetary-scale plate tectonics. Plate tectonic processes are a key part of our Earth history and we will see in later chapters how they have helped to regulate our climate, shape the evolution of life and biodiversity, and influence the distribution of mineral resources and precious metals.

Plate tectonics is one of the clearest expressions of Earth as an active planet. It has become increasingly clear that processes taking place deep in the planetary interior influence landscapes and life on the surface in multiple ways, including the very air we breathe.

# CHAPTER 10

# The Air We Breathe

All animals breathe to get oxygen into their bodies. The ladybird crawling across my study window, my cat sitting in a box behind me. Oxygen helps them turn food into energy. It helps them to grow and reproduce. Every tissue and cell in your body needs a constant supply of oxygen to function effectively. If the supply stops, irreversible tissue damage can occur and major organs, especially the heart and brain, can quickly fail. From sperm whales to earthworms, oxygen is essential for the existence of all complex animals.

The air we breathe is a mixture of invisible gases. Two of them, nitrogen and oxygen, make up more than 99 per cent of its volume. Earth's atmosphere is dominated by nitrogen (78 per cent) and oxygen (21 per cent) with some argon (0.93 per cent). The remainder (about 0.04 per cent) is a mix of trace gases including neon, hydrogen, helium and krypton, as well as the greenhouse gases carbon dioxide, methane, ozone and nitrous oxide.

Nitrogen (N) is also crucial for life on Earth. It is a key building block of DNA and essential for plant growth. Nitrogen in the atmosphere only becomes useful for plants following a process called fixation. This involves bacteria in the soil converting nitrogen gas into forms that plants can accept through their roots to make amino acids and proteins. The cycling of nitrogen through ecosystems is crucial for maintaining healthy plants and crops. Farmers increase crop yields with nitrogen-rich fertilisers.

Earth is the only planet in our Solar System with an oxygen-rich atmosphere. There may be planets in distant solar systems with atmospheric oxygen, but we have not yet discovered them. Oxygen is one of the critical ingredients that makes our world habitable and sustains complex life, but it was either absent or present in only tiny quantities for much of the first 1.5 billion years or so of Earth history.

The mix of gases in the atmosphere is not constant and scientists have used some ingenious methods to study the history of the atmosphere and its changing composition. We have analysed the air trapped inside centuries-old bottles of French wine and the ice age air preserved in frozen bubbles from ice cores drilled to depths of more than 2 kilometres into the ice sheets of Antarctica and Greenland (Chapter 33). Some of the Antarctic ice cores stretch back almost a million years, but they do not extend far enough to record the momentous changes that influenced the evolution of complex animals and plants. We must look to Earth's oldest rocks to find clues about the early atmosphere.

Active volcanoes emit gases constantly both during and between eruptions (Chapter 19). This is known as *volcanic degassing*, and this process has influenced the composition of the atmosphere from the earliest times. Earth's early atmosphere was dominated by nitrogen and carbon dioxide with some methane, water vapour and hydrogen. These gases were released from cooling magma and volcanic vents. Some of these gases would have been broken down by intense ultraviolet radiation from the young Sun. Oxygen was not present in the Hadean atmosphere – it is a highly reactive element, so in early Earth history it was usually bound up in compounds like water ($H_2O$), carbon dioxide ($CO_2$) and various iron oxides. This atmosphere would have been toxic to modern life.

*Insolation* is the amount of solar radiation that reaches the Earth. During the Archean Eon (4.0 to 2.5 billion years ago) the Sun was 25 to 30 per cent weaker than today. With such a weak Sun, Earth should have been locked in a permanent ice age, but the geological record of the Archean shows otherwise with the presence of oceans, flowing water and sedimentary rocks. This puzzle is known as the 'faint young Sun paradox' and has generated much debate.

How can we explain this? Several factors may have been in play to keep the young Earth warm, but most geologists agree that greenhouse gases played a key role in keeping a global ice age at bay. For most of the Archean, it seems that Earth's atmosphere provided enough *insulation* to offset weaker insolation. The Archean atmosphere contained a similar mix of gases to today, but they were present in quite different proportions and there is much uncertainty about when oxygen first appeared. Carbon dioxide, methane and water vapour were present in higher concentrations than today, but we cannot be precise about the percentages. Methane ($CH_4$) may have been the most important greenhouse gas at this time – it is a much more effective trapper of heat than carbon dioxide.

Life in an oxygen-free world was all about microbes. Archean life was dominated by an ancient group of single-celled, ocean-dwelling microbes called cyanobacteria, also known as blue-green algae. These tiny bioengineers get their energy via photosynthesis and eventually evolved the ability to produce free oxygen (oxygen gas that exists in the atmosphere that is not chemically bound to other elements). These organisms contain chlorophyll, a pigment that captures sunlight and starts the reaction. Cyanobacteria carried out this world-transforming activity by converting sunlight into chemical energy in the presence of water through a series of reactions that transform carbon dioxide ($CO_2$) into organic matter (carbon C) while releasing free oxygen ($O_2$) as a waste product. This process is known as *oxygenic photosynthesis*. This innovation was a game-changer for life on Earth. Over millions of years this process led to the oxygenation of Earth's atmosphere. Some of the oxygen in the atmosphere was converted into ozone ($O_3$), which helped to shield the Earth's surface and lifeforms from harmful ultraviolet solar radiation. When an ozone layer was in place, new species of microbes could emerge.

Around 2.4 billion years ago atmospheric oxygen rose rapidly in what is called the Great Oxidation Event (GOE). Organisms that could not adapt to oxygen-rich ecosystems became extinct or were forced to remain in specialised oxygen-starved niches. It would be difficult to overstate the importance of this new atmosphere of breathable air. The emergence of free oxygen as a significant component of the atmosphere transformed planet Earth, paving the way for the evolution of complex life in the oceans and on land. Before the GOE, any free oxygen produced was quickly absorbed by iron and other materials.

Before photosynthesis began oxygenating the atmosphere, large parts of the ocean lacked oxygen and were rich in hydrogen sulphide ($H_2S$). Some of the Earth's earliest microbes were sulphate-reducing bacteria that consumed organic matter on the ocean floor, releasing $H_2S$ as by-product. Unlike today's oceans, which are mostly oxygenated, Proterozoic oceans had vast zones where oxygen was absent, and sulphide was abundant. Hydrogen sulphide is toxic to most complex organisms, so these conditions limited the spread and diversification of oxygen-dependent organisms for most of the Proterozoic Eon.

When atmospheric oxygen increased steadily during the GOE, Earth shifted from a dominantly anaerobic (without oxygen) environment to an aerobic (with oxygen) environment. Anaerobic environments include organic-rich muds that accumulate on the floor of deep lakes, the guts of certain animals and hydrothermal vents on the deep ocean floor. These locations are not devoid of life, but they typically only support single-celled lifeforms such as bacteria. They often have low biodiversity.

The Archean and Proterozoic oceans created the iron-rich rocks that support today's iron and steel industry and almost every aspect of modern life. These oceans were rich in dissolved iron. The free oxygen released by cyanobacteria oxidised ferrous iron to ferric iron, which precipitated out of ocean water to form insoluble iron oxides that built up on the ocean floor. These rocks comprise thousands of alternating bands of iron-rich layers and layers of silica washed in by rivers. These are known as banded iron formations – they began forming before the GOE and continued until about 1.8 billion years ago when the ocean's iron supply was exhausted, and levels of free oxygen rose in both the ocean and atmosphere. With less dissolved iron left to react, banded iron formation slowed and eventually stopped. Banded iron formations are present on every continent and can be several hundreds of metres in thickness. Some of the largest iron-ore mines in the world exploit these ancient, banded rocks.

Physical geographers and geologists use the term subaerial (literally 'under the air') to describe processes that take place at or close to the Earth's surface. Subaerial weathering involves rocks and sediments exposed to Earth's atmosphere. The geological record changes profoundly after the GOE. The later record includes huge thicknesses of red sandstones packed with iron-rich minerals that have literally rusted in the presence of oxygen. The bright red sandstones of the Permian and Triassic periods are classic examples. The great civilisations of the Mediterranean world harvested olives, grapes and wheat from *terra rossa* soils rich in iron oxides. They still do.

If you walk through the elegant green parks of the National Mall in Washington DC from the Capitol Building towards the Washington Monument, you pass America's most iconic red sandstone building. The Smithsonian Institution Building (known as The Castle) is a striking Gothic Revival statement completed in 1855. The building stone is Upper Triassic sandstone, which sticks out like a bright red sore thumb on the National Mall against the cool white marble facades of the neoclassical government buildings. In the Triassic Period vast swathes of the Pangaea supercontinent were hot desert and semi-arid landscapes with some seasonal rainfall. This climate favoured the oxidation of iron.

Carbon dioxide ($CO_2$) accounts for only 0.04 per cent of today's atmosphere by volume but makes a vital contribution to life on

Earth out of all proportion to its tiny concentration. Without the greenhouse effect, Earth's temperature would be well below freezing and our planet would be in a permanent ice age (Chapter 17). The natural mean surface temperature of our planet today is just above 15°C because of the role carbon dioxide plays in regulating the Earth's energy balance. Without carbon dioxide, the natural greenhouse would collapse. Some estimates suggest the mean surface temperature could plummet by 35°C.

NASA has sent probes into space to study the composition of distant atmospheres. They show remarkable variety and remind us of the special conditions on Earth that make our world habitable. Titan is Saturn's largest moon and the only moon in the Solar System with a dense atmosphere. Its atmosphere is more than 98 per cent nitrogen. Venus has a very thick atmosphere dominated by carbon dioxide with yellow clouds of sulphuric acid: runaway greenhouse warming has created the hottest planetary surface in our Solar System with mean surface temperatures over 450°C and hardly any temperature difference between day and night.

The Moon has a very thin atmosphere called an exosphere. There is no oxygen or nitrogen; rather it is composed of helium, argon, neon, ammonia, methane and carbon dioxide. In the cold lunar night, without energy from the Sun, this exosphere collapses and falls to the Moon's surface. With no greenhouse effect to trap heat, the surface temperature on the Moon swings dramatically between day and night: in full sunshine, lunar temperatures can soar to over 125°C while in darkness the temperatures in lunar craters can fall to −173°C, producing a surface temperature range of almost 300°C!

The Earth's thick atmosphere also protects us from most meteorite impacts. Any meteorite smaller than about 25 metres across will burn up well before it gets close to the ground because of the extreme heat generated by friction during its high-speed passage through the atmosphere.

The mix of gases in the air we breathe is intimately linked to the geological and biological history of planet Earth. Understanding the origin and build-up of oxygen in the atmosphere and oceans is central to the study of the evolution of life. It also provides valuable context for the climate of the present day. The natural greenhouse effect is an essential element in making our world habitable, but you can get too much of a good thing. The geological record warns us that Earth can shift to super greenhouse conditions. The burning of fossil fuels is strengthening Earth's greenhouse effect and warming our planet so quickly that its habitability for humans may be under threat. We will explore this human-induced greenhouse warming in Chapter 37.

# CHAPTER 11

# The First Life on Earth

A fossil is a remnant or impression of a past lifeform that has been preserved in the rock record. We can think of the fossil record as a very thick book. The chapters of this book are scattered across seven continents and most of the pages are missing. Its readers are palaeontologists, who draw on expertise from many fields, primarily geology and biology, to reconstruct past ecosystems and the history of life. Some palaeontologists would argue that the most interesting part of the story is in the last few chapters, when complex life appears, and we see insects, fishes, amphibians, reptiles, birds, mammals and the like, all the way to the 8.7 million and more species of animals and plants living today. We will get to that part of the story later. But in order to flip back to the first pages to establish *when* the first organic molecules emerged – the origins of life itself – we must find the oldest fossils. Locating and then reading life's opening chapter is far from straightforward.

As we have seen, Earth's oldest rocks have been subjected to such intense heat and pressure that their original structure and mineralogy have been transformed. If the earliest lifeforms were single-celled algae, what is the chance of finding microscopic blobs of slime in the oldest metamorphic rocks that have been baked and squeezed for billions of years? The search for the earliest lifeforms has produced some ingenious research and a good deal of controversy.

It is hard to imagine our world without life, but we can divide Earth history into a time with life and a time without. Palaeontologists have scoured Archean rocks for ever-older fossils to establish when life began and to build a picture of the first ecosystems. This is one of the greatest challenges of Earth history. While the earliest fossil record is extremely sparse and difficult to read, under certain conditions it is possible to identify simple lifeforms from deep time. Some microbes lived in vast colonies where they combined with mineral grains to build robust structures that have survived in the rock record.

The earliest undisputed fossils are about 3.5 billion years old. They come from the Pilbara region of Western Australia where the rocks of the Dresser Formation were laid down in the Archean Eon around 3.48 billion years ago. These rocks are important because they contain stromatolites – the oldest *undisputed* fossil evidence of early life. Stromatolites are dome-like structures built by colonies of tiny sticky microbes that thrived in shallow tropical coastal waters. They glued sediment grains together to form layers, and these microbial mats built up layer by layer like a stack of thin pancakes – stromatolite means 'layered rock'. For a billion years of the Archean, life amounted to not much more than this: mats of slimy microbes living in shallow salty water.

We know quite a lot about stromatolites because they still exist today – a living example of one of Earth's earliest ecosystems. Southwest of the Pilbara, almost at the westernmost point of the Australian continent, Shark Bay attracts some 100,000 tourists every year who come to see the dolphins, turtles and dugongs – and the stromatolites. The Hamelin Pool Marine Nature Reserve is a UNESCO World Heritage Site that contains the world's largest and most diverse expanse of living stromatolites. Australia boasts the oldest and youngest stromatolites – a lineage that extends across the depths of geological time. Colonies of microbes have produced the same multi-layered dome-shaped mounds in ancient and modern oceans for more than 3.5 billion years.

What came before them? The oldest rocks formed on Earth are about 4 billion years old – that places a physical limit on the length of the fossil record. In the absence of recognisable fossils, geologists search for chemical traces in ancient rocks that point to the presence of organic matter and lifeforms. These are known as biosignatures. While this is only *indirect* evidence of early life, it can help to build a picture of the conditions under which life emerged.

We have already seen that Greenland has some of the world's oldest rocks. Early Archean rocks in western Greenland contain graphite, which is the crystalline form of carbon. In 2014 a group of scientists showed that some of these graphite grains displayed structures that are known to form when organic compounds are subjected to metamorphism. These ancient Greenlandic rocks do not contain fossils, but they contain a form of graphite that represents possible traces of early marine life from at least 3.7 billion years ago. This conclusion is highly controversial, but the researchers claim this is the earliest known evidence of life preserved in a discrete layer of rock of the same age.

We have looked at zircon crystals in Chapter 7. This remarkable mineral is so durable that crystals of zircon can be recycled by erosion many times without any change in their composition. The oldest zircons from the Jack Hills rocks in Australia are 4.4 billion years old – they date to a time just after the formation of our planet. These tiny time capsules allow us a glimpse into the environments of the Hadean Eon. Researchers have found graphite in a zircon crystal that is 4.1 billion years old and have argued that, like the example from Greenland, it is also derived from once-living matter. This interpretation is also strongly disputed.

Arguments about the age of the oldest fossils and how we interpret traces of ancient graphite will continue, but this is certainly an exciting time for the study of early life. New sites are being explored and new scientific techniques for identifying traces of ancient life are being evaluated. It is important to note that, from a collection of over 10,000 Jack Hills zircons, just one crystal has been shown to contain graphite. This tiny crystal has provided a possible tantalising glimpse of some of the first life on this planet. It may well be the earliest known evidence of life in the Universe.

Much of this work is controversial – an earlier study made similar claims about zircon crystals that were later found to have been contaminated with graphite in the laboratory during sample preparation. We certainly need more data, but there is a growing body of evidence pointing to Earth becoming amenable to life much more quickly than previously thought. This is a major shift in our understanding of early Earth history. If the dating is accurate and the interpretations are sound, the most recent work on the Jack Hills zircons shows that a primitive biosphere may have emerged by 4.1 billion years ago. The Hadean was so named because it was a time of boiling magma oceans and meteorite bombardment. The latest work suggests that at least the latter part of the Hadean may have been rather more temperate and habitable than hellish.

We do not yet have undisputed fossil evidence that predates the oldest stromatolites, but it is thought that methane-producing bacteria emerged some 3.8 billion years ago. Methane is a potent greenhouse gas, and these microbes may well have helped keep Earth's early climate temperate by strengthening the greenhouse effect at a time when the Sun was weaker than today. These early Archean microbes were the first lifeforms to change the global climate. A simple biosphere was up and running.

While science has made great strides – give or take a couple of 100 million years – tackling the question of *when* primitive life emerged on Earth, there is still much debate about *where* the tree of life began. What environment provided the key mix of ingredients to seed the first life and what processes were involved? Where did the life-creating energy come from?

The two environments that have been proposed as host for the origin of life on Earth could not be more different. One group of scientists has focused on the deep ocean floor at mid-ocean ridges, where energy comes from the Earth's interior and hot gases and mineral-rich liquids seep out from the mantle. Another group favours a shallow lake or coastal lagoon setting drenched in bright tropical sunlight. This latter group argues that warm nutrient-rich saline waters fuelled by light from the sun may have been the cradle of life. Did the first lifeforms appear in the murky darkness of the deep ocean or were they bathed in tropical sunshine? Laboratory experiments have attempted to simulate the chemical and biological processes in these environments, but we just do not know how life emerged.

Whatever the setting, geological processes were fundamental in creating favourable conditions for the emergence of life. The earliest microbes in the Archean Eon lived in an oxygen-free world. Instead of using sunlight for energy they took chemical energy from the environment, powering themselves by exploiting simple chemical reactions. They evolved metabolisms that used hydrogen, methane, sulphur or iron. These microbes probably emerged in or close to hydrothermal vents on the ocean floor and helped to shape the composition of the early atmosphere and the chemistry of the oceans.

The weathering of the continents flushed nutrients into the oceans that allowed cyanobacteria to thrive. The biosphere emerged as a key player in the Earth system during the Archean. We have already seen how these microbes took energy from sunlight and used water as fuel for oxygen generation. As life developed on Earth those lifeforms began to change the planet. They would eventually create an oxygen-rich atmosphere and pave the way for the emergence of complex life. But this took a very long time. While simple single-celled life emerged quickly in the early Archean, it would be another 3 billion years or so before the first animals appeared.

The period in the Proterozoic Eon between 1.8 and 0.8 billion years ago has been dubbed the 'boring billion' because it was initially thought that nothing of note happened. Nutrient supply in the oceans was limited and atmospheric oxygen levels stubbornly flat-lined for much of this interval. This evolutionary standstill may have delayed the evolution of complex life. More recent research has argued that this period may not have been quite so dull after all and that the environmental stresses caused by an extended period of low nutrient availability may have triggered important evolutionary innovations. We now know that this period was actually quite eventful, with key evolutionary changes including the appearance of multicellular lifeforms and sexual reproduction.

A novel approach to exploring the history of life involves the use of modern genetic data and mathematical models to calculate the speed of evolution. By working backwards along the branches of the tree of life, with a few key dates from the fossil record to estimate the rates of evolutionary change, you eventually come to a common ancestor – the single-celled microbe from which all of today's living beings on Earth have evolved. This is known as the *last universal common ancestor*, or LUCA.

It is important to note that LUCA is *not* the origin of life – that must have been sometime earlier. A recent study that used this genetic clock suggested that LUCA lived about 3.9 billion years ago. This is intriguing because it lends support to the idea that life emerged in the Hadean and before the chaotic period known as the Late Heavy Bombardment (Chapter 2), when Earth was repeatedly struck by large meteorites. The conventional view held that life could not survive that bombardment. If the LUCA calculations are correct, the first organisms would have appeared even earlier and very likely before 4 billion years ago. This is an intriguing result because, like the graphite in the Jack Hills zircons, it adds weight to the notion that Earth become habitable much earlier than previously thought.

Various approaches are allowing us to peer into Earth's early history. A remarkable picture is emerging of a planet with oceans of water that became habitable – at least for the most primitive lifeforms – soon after its formation. But we still have many more questions than answers. We just don't know how and when life emerged on Earth. Once the key ingredients and environments were in place, it is even possible that life emerged more than once.

All of this raises fascinating questions around whether these processes could take place on other planets. We don't know if there has ever been life on Mars, but we know it used to be a habitable planet. Its surface contains vast canyons and ancient lake basins – graphic evidence of an active hydrological cycle and a climate that once supported lakes and rivers. These conditions may have lasted for more than a billion years. Why did Mars lose its liquid water and habitability, but Earth did not? We may be close to finding evidence of ancient primitive life on Mars. We need more rock samples from Mars before we can tackle this question but the fact that we are asking it shows that, once established, planetary habitability is not guaranteed.

# CHAPTER 12

# Earth's Magnetic Shield

Writing in the first century AD in his *Naturalis Historia*, Pliny the Elder (AD 23/24–79) described a strange hillside where the rocks attracted iron. While tending his sheep, the shepherd Magnes found that the nails of his shoes and the iron tip of his staff stuck fast against a rock. He was the first to observe the phenomenon of magnetism. We do not know if Magnes was entirely mythical, and the story has no doubt been adorned over the centuries, but he will forever be associated with the mineral magnetite and the wonders of magnetism.

Scholars in ancient Greece and China were aware of magnetism even if they did not comprehend its relation to Earth's magnetic field. By the seventh century AD Chinese scholars had worked out how to make iron needles magnetic by rubbing them with magnetite. Much later these needle magnets were floated in water and suspended from fine silk threads and used as navigation devices at sea.

In England in 1600, William Gilbert published his famous book *De Magnete*, where he set out a theory of Earth's magnetism. A gifted designer of experiments, Gilbert was first to conclude that magnets align themselves in a north–south direction because the Earth itself is a magnet. He also debunked medieval superstition that magnetism could be destroyed by garlic and that a magnet could detect an unfaithful wife! He was physician to Queen Elizabeth I in the final years of her life.

Today magnets of varying complexity have hundreds of practical uses in industry, medicine and around the home, from separating metals in recycling plants to keeping holiday photos fixed to the fridge. Magnets are used to store data in computers and magnetic resonance imaging uses strong magnetic fields combined with radio waves to produce detailed images of the human body. Even though magnetism is literally everywhere, it is rarely appreciated how the invisible magnetic field that surrounds our world plays a vital role in sustaining life on Earth. Without a magnetic field, Earth's atmosphere would be degraded by harmful radiation and gradually stripped away by the solar wind, with potentially catastrophic consequences for all life.

Earth has a strong magnetic field but other rocky planets like Venus and Mars do not. Deep below the Earth's surface the liquid outer core generates Earth's magnetic field. Driven by heat trapped in the core flowing out to the mantle, large-scale convection processes and the spinning Earth keep this ocean of molten iron and nickel in constant motion. As this metallic mass churns, it induces a magnetic field through a self-exciting dynamo process that converts the movement of an electrically conductive fluid into electromagnetic energy. This giant dynamo creates a geomagnetic field that extends around the entire planet and atmosphere. Venus and Mars do not create this planetary-scale dynamo effect. Their cores do not swirl vigorously enough because they are not losing heat fast enough to their mantles. The Moon lacks a power source to generate a magnetic field because its core is too small or too cool or both.

When observed from space and at the Earth's surface, the magnetic field looks like that produced by a simple bar magnet; the field forms two poles with opposite positive and negative polarities. The field can be detected using a magnetic compass with a small bar magnet needle that can rotate. The compass needle points in the direction of the Earth's magnetic field. If you put a small bar magnet on a piece of paper and sprinkle iron filings around it, the iron filings become induced magnets and line up with the field. You can draw the directional lines of the field on the paper. The Earth's magnetic field has broadly the same pattern but in three dimensions.

The extent of the magnetic field around a planetary body is known as the magnetosphere. The Earth's magnetic field extends outwards from the core into space where it meets the solar wind, the stream of charged particles emanating from the Sun. Earth's magnetosphere is highly dynamic and responds dramatically to changes in solar activity. The magnetosphere protects most of the Earth's surface and atmosphere from cosmic rays from deep space and from bombardment by harmful solar radiation. It deflects harmful charged particles from the Sun, forcing them to travel around the planet. Some of these charged particles follow magnetic field lines down towards the North and South Poles and this is why auroras happen mostly in polar regions – the field lines bend down and let the particles enter the upper atmosphere where they collide with gases in the atmosphere. These collisions cause the spectacular glowing colours of the ribbons, arcs and swirls of the aurorae. The northern lights (*aurora borealis*) can been seen most frequently near the Arctic Circle in places like Alaska, northern Canada, Iceland, Norway, Sweden and Finland. The southern version (*aurora australis*) can be seen in Antarctica or southern parts of New Zealand and Australia.

Earth's magnetosphere is comet-shaped. It is blasted by the solar wind and squashed on the sunward side so that the magnetosphere extends towards the Sun for a distance of some six to ten times Earth's radius. On the side of the Earth in darkness, it has a spectacularly long tail that extends up to a thousand times Earth's radius because it is dragged outwards by the solar wind.

Earth has the strongest magnetosphere of all the planets in our inner Solar System. It is one of the key ingredients that makes our world habitable. We only have to look to Mars, which has no magnetic field – the Martian atmosphere has been largely stripped away. Life on Earth's surface is protected by the magnetic environment generated in our planet's deep interior. And, since life on Earth has evolved within this magnetic force field, it also influences the behaviour of the animals that call our planet home.

It seems incredible that creatures such as Pacific salmon and sea turtles can find their way back to where they were born across thousands of kilometres of open ocean when strong currents could easily send them off course. Loggerhead sea turtles (*Caretta caretta*) undertake one of the longest migrations in the animal kingdom. Each North Atlantic loggerhead hatchling born on Florida beaches begins a 13,000-kilometre solo journey around the North Atlantic basin that takes in the coastal waters of West Africa. First, the baby turtles have to navigate the perilous coastal waters which are targeted by predators, and it is vital for the survival of the species that they swim to the open ocean. They then spend years at sea in the North Atlantic gyre, a huge clockwise circulation in the widest section of the ocean, before returning to the beach where they were born to nest and lay their eggs. Research published in *Science* in 2001 by marine biologist Kenneth Lohmann (b. 1959) and colleagues at the University of North Carolina showed that young sea turtles exploit regional magnetic fields as navigational markers. These turtles have an inbuilt GPS guided by magnetic field intensity and magnetic inclination angle. Cues from the Earth's magnetic field guide these turtles along their epic journey.

The notion that animals might use 'magnetic maps' for long-distance navigation was highly contentious just a couple of decades ago. It is now well established and an exciting research frontier in zoology. Indeed, Kenneth Lohmann has pointed out that since life arose in Earth's magnetic field, we should not be surprised that some animals have evolved the ability to use it to guide their movements. The most recent discoveries indicate that the magnetic field of their birth area is imprinted in the juveniles of some animal species and this information helps them to navigate their return as adults.

NASA and other space agencies invest considerable resources in researching the behaviour of Earth's magnetosphere and how it interacts with space weather and threats from the Sun. Solar storms send out huge clouds of electrically charged particles that can travel deep into space. Some of these particles will reach the Earth's upper atmosphere. Since most spacecraft, satellites and global communication systems operate within the magnetosphere, a better understanding of the processes and potential hazards in this zone is of huge commercial and strategic importance.

The strength of the Earth's magnetic field fluctuates over time because the forces in the deep interior that create the field are constantly changing. One of its most remarkable features is its propensity to flip. From time to time our planet's magnetic field reverses completely so that magnetic north becomes south. The cause and impacts of these reversals are not fully understood but we know they happen because they are written in the rocks. *Palaeomagnetism* is the study of the past behaviour of Earth's magnetic field through the analysis and interpretation of magnetic signals preserved in rocks and sediments.

Perhaps the most famous example of palaeomagnetic research relates to the alternating stripes of magnetic anomalies observed in the rocks of the deep sea floor either side of mid-ocean ridges. As molten basalt emerges from the mantle to create new ocean crust on either side of the ridge, the direction of the Earth's magnetic field is recorded in the minerals as they cool.

Ocean crust contains significant amounts of magnetite and the tiny magnets in the iron-bearing particles in the minerals align with the field as the lava cools. The rocks are classified as having normal polarity (like today) or reversed polarity when the field has flipped. These magnetic anomalies form parallel stripes on either side of the ridge. The basalt conveyor belt moving in both directions away from the mid-ocean ridges proved to be the perfect setting to record long-term changes in the Earth's magnetic field. Not only did these observations prove that the magnetic field repeatedly reversed its polarity, they also provided the hard empirical data that were so critical in the 1960s to validate the theory of sea floor spreading (Chapter 9).

The geological record shows hundreds of magnetic reversals and this is a very clear demonstration that the Earth's magnetic field is highly dynamic. On average a reversal happens about three times per million years, but they are not regularly spaced in time. The last full magnetic reversal took place 780,000 years ago in the middle of the Pleistocene ice age and is known as the Brunhes–Matuyama reversal. Such dramatic changes to the Earth's magnetic field are of great interest because they reflect big shifts in behaviour deep in the Earth's interior.

We can even see this dynamic magnetic interplay at work today. Were you to dig out your compass right now, you'd see the needle pointing towards the magnetic north pole. But this is not a fixed point. Because of the dynamic processes within the Earth's molten core, magnetic north constantly wobbles around. When it was first located in 1831 by the navy commander and polar explorer James Clark Ross (1800–1862), magnetic north was located in the Nunavut territory of Arctic Canada. Since the last decade of the twentieth century it has been on the move in the direction of Russia at a rate of some 50 to 60 kilometres each year.

How old is Earth's magnetic field? This is an important question because the creation of the field would have provided conditions that were more favourable for the formation and preservation of an atmosphere and thus for the evolution of the first lifeforms. Zircon crystals from the Jack Hills region of Western Australia have provided important clues about the age of the magnetic field. Using an extremely sensitive magnetometer, geophysicists have detected faint magnetic signals in the iron-bearing materials inside the 4-billion-year-old zircons. These magnetic signals tell us about the strength and direction of Earth's magnetic field when the zircon crystals formed. They suggest the magnetic field and the processes that generate it have been in operation since at least the end of the Hadean Eon, even if the field was weaker and the magnetosphere much smaller than at present. It must be pointed out, however, that quite a large contingent of the palaeomagnetic community is rather sceptical about these results. They argue that the oldest good evidence for the presence of a magnetic field is at 3.5 billion years ago within rocks from the Barberton Greenstone Belt in South Africa.

Magnetic signals preserved in ancient rocks have been used to reconstruct the changing geography of the continents. They also provided key information that led to geologists uncovering evidence for the most severe ice ages in Earth history.

# CHAPTER 13

# Snowball Earth

The tiny island settlement of Port Askaig on Islay in Scotland's Inner Hebrides is famous for its wild Atlantic weather and single malt whisky. It is also the site of an extraordinary nineteenth-century geological discovery. Despite little formal schooling, the self-taught Scottish geologist James Thomson (1823–1900) was an astute observer of nature who became an authority on the ancient rocks of his homeland. In 1871 he published a remarkably insightful short paper about a suite of glacial features – including far-travelled granite boulders – that he recognised in Precambrian rocks near Port Askaig. These ancient glacial traces formed during the most severe and prolonged ice age in Earth history. Thomson's paper was the first piece in a global geological jigsaw puzzle that is still being pieced together today.

In the Cryogenian Period (720 to 635 million years ago), the Earth plunged into prolonged and extreme deep freezes when ice sheets developed in the tropics. Indeed, the latest scientific dating tells us that Earth experienced extreme glacial conditions for much of the Cryogenian's 85 million years. Imagine our planet completely entombed in ice, a wasteland of glaciers surrounded by oceans frozen to a depth of more than 1,000 metres. The idea that this happened is now known as the Snowball Earth hypothesis. It's a hypothesis because the full extent of the deep freeze is disputed. It argues that our world – from pole to equator – experienced extreme ice age conditions for tens of millions of years before the development of complex plants and animals, and at a time when global geography was vastly different from today.

'Precambrian' is the term that was first introduced by geologists to cover the great span of time from the formation of the Earth to the evolution of complex life. It accounts for about 88 per cent of the geological timescale from 4.567 billion to 538.8 million years ago. Some consider it a rather old-fashioned term, but it's still widely used. Geologists discovered evidence of 'Precambrian glaciation' in various parts of the world throughout the twentieth century. When the deposits left by glaciers become hardened into solid rocks they are known as *tillites*, and Precambrian tillites were found on every continent, from China to Brazil, encouraging geologists to speculate about the nature of this ancient ice age and how it might fit into the story of life on Earth.

Many tillites have been found in locations that would have been in tropical latitudes at sea level during the Cryogenian Period. This raises all kinds of questions about what the climate was like at this time because modern glaciers only extend to sea level in the polar regions. Today only small glaciers are present above the snowline on the very highest tropical mountains such as Kilimanjaro and Mount Kenya, and even these are receding rapidly because of global warming.

Some of the best evidence for Precambrian glaciation is recorded in the rocks of South Australia. Douglas Mawson (1882–1958) was part of Ernest Shackleton's 1907–9 Antarctic expedition. Like Shackleton, he is famous for heroic polar exploration, but he was also a pioneer of Precambrian glaciation. Having observed the work of Antarctic glaciers, his geological research at the University of Adelaide led to the discovery of glacial beds that extend for over

1,500 kilometres in South Australia. The great thickness of these rocks indicated that glacial conditions had persisted for an immense but unknown length of time.

Much later, Brian Harland (1917–2003) was the first geologist to pull together all the Precambrian glacial evidence and put forward the bold idea that global-scale glaciation had taken place early in Earth history. Harland was based in Cambridge but spent extended periods in cold climates close to glaciers. He was an expert in recognising in very ancient rocks the traces left by glacial ice. Beginning in 1938, he spent forty-three field seasons studying the geology of the Arctic archipelago of Svalbard.

In 1964 Harland concluded that a Precambrian ice age was sufficiently extreme to form marine tillites in the tropics. Making the case for extensive glaciation in the tropics was an extremely radical idea at the time. Harland formed these ideas in the 1950s and 1960s when the theory of continental drift was becoming widely accepted. He was an early supporter of the theory, not least because it might help explain the widespread presence of glacial deposits in the low latitudes. Harland's work stimulated much debate and provided key data for the Snowball Earth theory that emerged in the 1990s, although he remained sceptical of the more extreme scenarios proposed by some of its later exponents.

The cryosphere (from the Greek *kryos*, 'cold') is made up of all those parts of the Earth's surface and near surface where water is frozen. It includes snow, glaciers, ice sheets, permafrost, lake and sea ice and even hailstones – all forms of ice. Today's cryosphere is dominated by the huge ice sheets of Antarctica and Greenland, massed in the polar regions. Snowball Earth involves the cryosphere expanding to overwhelm the *entire* planet.

How could large ice sheets have extended over the tropics? What mechanism could send the Earth plummeting into a *global* ice age that would have made it look like a giant snowball from outer space? In the late 1960s, the brilliant Russian climate modeller Mikhail Budyko (1920–2001) proposed an explanation based on the ability of snow and ice to reflect the Sun's energy back into space. Snow and ice have a high *albedo* (from the Latin *albus*, meaning 'white'). The albedo of a surface is a measure of how effectively it reflects energy from the Sun.

In 1969 Budyko published one of the first theoretical studies of what is known as the snow and ice–albedo feedback mechanism. Because snow and ice reflect solar energy back into space, if the area covered by snow and ice increased, the Earth would cool and more snow would fall to feed glacier expansion. This would cool the climate further, and so on. From his calculations, Budyko postulated that once ice reached a latitude tipping point of 50° (north or south), sea ice and glaciers could extend to cover the Earth's surface with ice growth across the ocean surface and on land sustained by powerful feedback loops.

Budyko's calculations indicated that the albedo effect could promote a *runaway glaciation*. Budyko was originally thinking about these matters because there was a very real prospect of a nuclear winter during the Cold War. The worry was that nuclear blasts could inflict irreversible damage to the atmosphere, so he wanted to understand what impact this might have on Earth's climate. Advocates of Snowball Earth have cited Budyko's modelling as a realistic mechanism for triggering a global freeze.

It was the work on Precambrian tillites by Caltech professor of geobiology Joseph Kirschvink (b. 1953) that provided fundamental evidence in support of global glaciation. He was the first to coin the term Snowball Earth in 1992. Kirschvink is an expert on the magnetic properties of materials. As we saw in Chapter 12, the magnetic properties of rocks retain information about the strength and direction of the Earth's magnetic field from the time of their deposition. So, by studying the magnetic properties of a rock sample you can work out the approximate latitude of the site where it was deposited. Joe Kirschvink used magnetic data to show definitively that many Precambrian glacial sediments were laid down in the tropics and some were deposited at the equator. He argued that if large glaciers were present in equatorial latitudes, the entire planet would have been in a deep freeze. This is the key argument for a Snowball Earth.

Paul Hoffman (b. 1941) is a Harvard geologist and one of the leading advocates of the Snowball Earth hypothesis. He has gathered evidence on ancient glaciation from around the world. Some of the best evidence for Snowball Earth glaciers comes from the most unlikely places, including the Namibian desert. The latest data show that there were two main periods of Cryogenian glaciation – these are known as the Sturtian and Marinoan glaciations. Both are named after rock formations first observed by Douglas Mawson in South Australia before the First World War. The Sturtian glaciation dominated the Cryogenian, lasting some 57 million years from about 717 to 660 million years ago. The Marinoan glaciation came later and lasted for about 10 million years although dating control is less secure. Hoffman and others have argued that both were worldwide glaciations under Snowball Earth conditions. We now know that the rocks on Islay that James Thomson described over 150 years ago belong to the Sturtian glaciation.

The geological record shows that both glacial periods ended abruptly with ice-transported deposits immediately overlain by limestone and dolomite rocks known as 'cap-carbonates'. This association of tropical carbonate rocks sitting on ice age rocks without a major gap between them indicates a very unstable Cryogenian climate flipping rapidly from one extreme to another: a challenging environment for any lifeform.

There is no disputing that the Cryogenian saw the most extreme glacial conditions in Earth's history, but only the hardened Snowball Earthers argue that the planet was completely frozen over. Global geography was very different in the Neoproterozoic Era, with most land masses sitting in the tropics forming part of a supercontinent called Rodinia that broke apart during the Cryogenian. This geography produced a vastly different global climate system from that of today. The Sun was about 6 per cent weaker at this time. Ice and snow in these low latitudes could generate a powerful albedo feedback effect because more sunlight is received in these latitudes and more energy would therefore be reflected into space.

Climate modelling has produced a plausible mechanism to create a Snowball Earth, but how could Earth ever escape from such a deep freeze? What triggered a global thaw? The best explanation is greenhouse warming. Throughout the deep freeze, volcanoes would have continued to pump carbon dioxide and other gases into the atmosphere. Rock-weathering processes normally consume some of this carbon dioxide, which then becomes locked up in carbonate shells in the oceans (Chapter 17). But these processes were subdued during the Cryogenian glaciations, allowing carbon dioxide to build up to extremely high concentrations in the atmosphere. Earth eventually reached a super greenhouse state when warming triggered rapid melting of the ice, sending the ice–albedo feedback into reverse. Spring finally came after millions of years of winter.

Some Snowball Earthers argue that the oceans were largely sealed off, with thick sea ice covering their entire surface. In places the sea ice may have been 1,000 metres thick. Other models suggest the Earth did not completely freeze over and there was open water in the tropics throughout the Cryogenian. It is important to remember that the marine environment must have sustained some life throughout this period. These refuges may have been at mid-ocean ridges in the deep ocean where warmth and nutrients were supplied by tectonic processes.

The Cryogenian glaciations are particularly remarkable because ice ages are rare in Earth history – only a tiny fraction of the last 4.56 billion years was conducive to the formation of large ice sheets. The geological record contains evidence for an even older period of glaciation that took place some 2.2 billion years ago. Whether this was a planetary-scale deep freeze has not yet been established. The clearest and most abundant evidence for anything approaching Snowball Earth conditions comes from the Cryogenian Period.

The Snowball Earth hypothesis challenges conventional ideas about the extent of global change. We think of Earth in the 'Goldilocks Zone' as never too hot or too cold, but periods of intense and widespread low-latitude glaciation rapidly shifted to super-greenhouse conditions in the Cryogenian as global temperatures soared and limestones were deposited in warm tropical oceans. It is a fascinating period in Earth history because the cryosphere expanded to such an extent that it dominated all parts of the continents and oceans, yet life was able to survive in a world dominated by ice and extreme cold. The cryosphere would never be so dominant again. Before the Cryogenian, life on Earth was dominated by simple microbes. After the Cryogenian we see the emergence of a remarkable diversity of complex life in the oceans. We will explore this evolution in a later chapter.

# CHAPTER 14

# Where Did All the Water
# Come From?

Earth is a blue planet. Seventy per cent of our world is covered by oceans with an average depth of more than 3.6 kilometres. That is a huge volume of water. Earth has oceans, lakes, rivers and waterfalls – it is the only place in the Solar System where water is abundant as a liquid on the planetary surface. Even though water is so familiar and one of the defining features of our world, we still do not fully understand how our planet came to be so wet. Where did all this water come from and how old are the oceans? These questions are fundamental to our Earth history.

Most geoscientists agree that in the first few tens of millions of years after the Earth's formation, the planetary surface had no water or atmosphere. Earth formed in the inner Solar System where scorching temperatures were far too hot for water to exist in liquid form. We do not know precisely how long this state was maintained, but the surface of the young Earth was incandescently hot with oceans of bubbling magma and temperatures well over 2,000°C. This molten state was sustained by frequent bombardment by comets and asteroids. This early Hadean environment was hardly suitable for a liquid that vaporises at 100°C. So when was this super-heated surface replaced by an ocean of water? How do we make a blue planet?

An ocean of deep water does not form overnight, and we cannot be precise about when the first basins on Earth's surface began to fill with water. But we can get a good handle on the antiquity of the oceans by dating the oldest known rocks that were deposited in deep water. Ancient rocks in southwestern Greenland hold some clues. Not far from the capital Nuuk, there are rocks known as the Isua Supracrustal Belt that are 3.8 billion years old. These are some of the oldest rocks on Earth (Chapter 7). In places these rocks are very strongly metamorphosed, but they also include exposures where pillow lavas and banded iron formations have been identi-fied. Both were laid down in deep water.

Banded iron formations were deposited in distinctive layers when iron-rich water from the deep ocean mixed with a surface layer where the presence of oxygen allowed iron oxides to precipi-tate (Chapter 10). Pillow lavas form where lava emerges slowly into a body of water. The water cools the surface of the lava, which then crusts into an expanding pillow shape inflated by the steady injec-tion of lava. The bulbous lobes of lava break off when they attain a critical size. These can topple down the steep slopes at mid-ocean ridges and pile up like pillows stacked on a bed. These rocks in southwestern Greenland were laid down in an ancient ocean early in the Archean Eon. The dates we have from Greenland provide a *minimum* age for the formation of the deep ocean. The evidence points to the emergence of Earth's oceans in the Hadean Eon. For a while there may have been a global ocean before the first conti-nents formed.

Environmental clues stored in tiny zircon crystals from the Jack Hills region of Western Australia may record the earliest evidence of rain falling on the land. The chemical composition of the crys-tals indicates they formed in magmas mixed with fresh water about 4 billion years ago. This suggests that Earth's water cycle was up and running in the Hadean.

Even though there is uncertainty about when the first extensive areas of dry land were in place, we have solid evidence from the geological record to show that deep oceans of water have been present for much of Earth's history. But where did this water come from? There is a very lively scientific debate about the origin of Earth's water and there are two rival theories that could not be more different. One suggests that Earth's water was delivered from space by meteorites and icy comets. The second theory argues that Earth's water has been present from the time of our planet's formation. This water was trapped inside the dust and rocks that accreted to form the Earth, then sweated out from minerals in the mantle and released to the surface as vapour via volcanoes.

The abundance of water deep in the Earth's mantle has been confirmed from an unusual source: diamonds. They are the hardest natural mineral with a very rigid crystal structure. Diamonds are also incredibly old. Most natural diamonds formed between 3.5 billion and 1 billion years ago under huge pressure and temperature deep in the mantle, and were then ejected by volcanic eruptions and incorporated into igneous rocks known as *kimberlites*. A jeweller examining a diamond seeks purity and perfection. Imperfections, called *inclusions*, make the diamonds less valuable. But Earth scientists like inclusions – they prefer their diamonds to be dirty. This is because during their formation diamonds capture information about the chemistry and physics of the mantle environment and then bring it unaltered to the Earth's surface. By studying the inclusions in ultradeep diamonds we can peer into Earth's mysterious deep interior.

A study of ten diamonds collected from several parts of Africa and China found microscopic inclusions – just a few microns across – filled with a crystalline form of water known as Ice VII. This form of water has a distinctive crystalline structure that is only created under high pressure. To give you an idea, the average atmospheric pressure on Earth at sea level is 1 atmosphere; Ice VII only forms when the pressure exceeds 30,000 atmospheres. Two of the diamonds with Ice VII inclusions contained structures that indicated they had originated in the lower mantle between 600 and

800 kilometres below the Earth's surface. This study provided the first evidence for the presence of water-rich fluids at such depths. It seems likely that most of the water in the Earth's oceans has been recycled at least once through the Earth's mantle over geological timescales. This is a startling discovery that reinforces how the workings of our planet – inside and out – are intimately connected.

It might seem counterintuitive to suggest the presence of water in the Earth's mantle when temperatures can exceed 3000°C, but water can be stored in various forms inside the lattices of rock minerals and the immense pressure in the mantle allows fluids to exist that would vaporise at the surface. So how much water is stored in the Earth's mantle? It is difficult to be precise about this figure, but most scientists agree that the volume of water in the mantle lies somewhere between the equivalent volume of water in today's oceans and ten times that amount. That is a huge store of water, but we are not talking about a huge subterranean reservoir. This water is dispersed through the mantle with some trapped inside minerals and some forming part of mantle fluids that flow under extreme pressure.

Between the lower and upper mantle, at a depth of 410 to 660 kilometres, the mantle transition zone is predicted to be a major store of water. We cannot sample this zone directly, but an elusive mineral known as *ringwoodite* is believed to be common at these depths. Olivine is the most common mineral in the upper mantle (which extends to a depth of about 670 kilometres below the Earth's crust) and ringwoodite is a high pressure version of olivine. Ringwoodite is important because it traps the ingredients for water within its mineral structure, holding hydrogen and oxygen atoms bound together in a form known as hydroxides. A tiny crystal of ringwoodite was found in a diamond from Brazil in 2014 – the first time this mineral had been seen in a terrestrial rock. Before this discovery ringwoodite had only ever been observed in meteorites or in laboratory simulations. If just 1 per cent of the mantle rock in the transition zone was $H_2O$, it would be equivalent to almost three times the volume of the world's oceans.

During the formation of Earth and the other rocky planets, temperatures in the outer Solar System were cold enough for water to freeze. This allowed the giant ice planets, Neptune and Uranus, to form. There is a growing body of evidence pointing to Earth's water being part of planetary formation, but we cannot yet rule out some contribution from comets and meteorites because we have only sampled a small range of meteorite types. Comets can be up to 80 per cent ice, and they originate in the outer Solar System, but the composition of the water they carry generally does not match the isotopic composition of Earth's oceans.

There are two main groups of meteorites that land on Earth. *Achondrites* have detached from larger rocky bodies that have been melted at some point in their history. This melting tends to strip them of water so these are bone-dry meteorites and cannot be the source of Earth's water. Another group of meteorites called *carbonaceous chondrites* can hold about 20 per cent water by weight. Although we are not yet able to quantify their contribution, the scarcity of these particular space rocks suggests it is unlikely that meteorite impacts could account for the great volume of water in the Earth's oceans.

$H_2O$ is the most well-known chemical formula; it shows that a molecule of water contains two atoms of hydrogen and one of oxygen. Hydrogen and oxygen are also the first and third most common elements in the universe, so we should not be surprised that water has been found in some unexpected places. Water ice is certainly present on all the rocky planets in our Solar System apart from Venus and there is a strong possibility that liquid water may be stable beneath the surface of the Moon. Mercury and the Moon have water ice stored in permanently shadowed craters that was delivered by comets. Mars has large masses of water ice in its polar regions, but lost its liquid water to space billions of years ago. The loss of its magnetic field, weaker gravity, erosion of the atmosphere and absence of plate tectonics may all have played a role. The presence of water oceans on its surface sets Earth apart from the other planets in our Solar System. Perhaps we should be asking how Earth has kept its water for so long. Plate tectonics may offer some clues.

There is a long-term component to the Earth's water balance that is rarely mentioned in textbooks. Water plays a role in lubricating plate tectonic processes, and over geological timescales Earth's water is cycled through the mantle. Water is lost from the oceans along subduction zones as slabs of crust with water-bearing sedimentary rocks are thrust downwards into the mantle (Chapter 9). This lost water is replaced at the surface as water vapour escapes along mid-ocean ridges and from subduction zone volcanoes.

Geoscientists have estimated the volume of these gains and losses and found them to be broadly comparable. This agrees with the geological record that shows that global sea level has not shifted dramatically (more than about 130 metres) over the last few billion years. Sea level tends to be lowest when supercontinents are formed and highest when the landmasses are fully dispersed, but the volume of the oceans has stayed fairly stable over this time period. Without the return of water to the mantle at subduction zones, sea level would have kept rising, submerging the continents. So plate tectonic processes are a key component of the long-term water balance of our planet, with lithosphere and hydrosphere in elegant equilibrium as part of a dynamic Earth system.

Earth's water cycle involves much more than the $H_2O$ that moves between the atmosphere, oceans, lakes and rivers. The geological water cycle extends hundreds of kilometres below our feet deep into Earth's interior, where there is a huge hidden store of water in the mantle that very likely exceeds the volume of the oceans at the Earth's surface. The latest research on the mantle has shown that we must now think about a water cycle for the *whole* Earth.

# CHAPTER 15

# The First Animals

What do a sperm whale, a bumble bee, a *Brontosaurus* and a trilobite have in common? They are all animals. But what is an animal? Think about that for a moment. If we want to explore the fossil record to establish when the animal kingdom began, we need to agree on the characteristics that define an animal. Being multi-cellular is not enough. An animal must have a body of many kinds of cells that perform specialist tasks. It must get energy by eating other organisms. It must breathe oxygen and be able to reproduce sexually. It must be able to move, sense its environment and react to stimuli. Animals are beautifully complex.

When did cells develop the ability to build eyes, lungs, brains, sexual organs and skeletons? Establishing when and how animals first evolved has proved to be challenging for palaeontologists because the fossil record is imperfect, and it is not always possible to identify these characteristics in incredibly old fossils, especially when creatures have no living relatives. Nonetheless, the search for the first ecosystems with animals is a crucial part of our Earth history and exciting discoveries regularly appear in the newspapers and on TV. The emergence of animals is a key milestone that leads directly to the evolution of our own species.

The chances of any creature entering the fossil record are vanishingly small. Some are eaten by scavengers; most decompose without trace. The fossil galleries in natural history museums are full of big bones and hard shells; these are the durable parts of long-dead animals that are most likely to be preserved in the geological record. In this way, the fossil record is inherently biased: the earliest animals emerged in the oceans and were soft-bodied, so they are always underrepresented. Animals like jellyfish are only preserved under special conditions, such as rapid burial by soft muds in deep water with a limited supply of oxygen to prevent decomposition.

Rocks containing fossils that are exceptionally well preserved are sometimes called *Lagerstätten* (from the German *Lager*, meaning 'storage', and *Stätte*, meaning 'place'). Such settings can preserve the precise imprints of delicate soft tissues if decomposition of the animal is slowed down for long enough to allow sedimentation to capture a detailed impression of the body, like a fingerprint in wet mud. In some settings the soft tissue may be replaced by minerals.

The fossil record tells us that animals have been absent for most of Earth history. While single-celled microbes dominated life for 2.5 billion years, all this changed dramatically in the Cambrian Period (538.8 to 486.9 million years ago). The Cambrian is the first period of the Palaeozoic Era. Palaeozoic means 'ancient life'. This era includes six geological periods spanning almost 300 million years: Cambrian, Ordovician, Silurian, Devonian, Carboniferous and Permian.

The Cambrian was a key turning point when a remarkably diverse array of complex multicellular organisms, including beautifully preserved soft-bodied animals, appear in the fossil record. The emergence of this spectacular marine menagerie, known as the *Cambrian explosion*, was geologically rapid. This extraordinary blossoming of life saw the evolution of complex marine ecosystems and the beginnings of the evolutionary arms race between predators and prey.

New means of propulsion emerged that meant animals could move further than ever before by swimming, sliding and scurrying across the ocean floor. For the first time creatures could also modify their physical environment by burrowing in sediment and taking elements from seawater to build hard shells and skeletons. The Cambrian Period saw the widespread appearance of animals with hard exoskeletons, such as trilobites, that facilitated their preservation. As it takes a lot of effort and resources to produce a hard exoskeleton, this innovation may have been a response to the emergence of the first predators. Some of the Cambrian beasts had body forms that are remarkably similar to many of today's arthropods, such as crabs, crayfish and millipedes, which account for about 80 per cent of living species in today's biosphere. This period has been the subject of intense study since the mid-nineteenth century and while many of these animals disappear from the fossil record, the Cambrian explosion laid the foundations for the diverse array of lifeforms we see today. It produced the major branches that make up the tree of life.

The Cambrian explosion is a key episode for understanding the evolution and diversification of complex multicellular organisms. Why did animal life suddenly diversify so spectacularly more than half a billion years ago in the Cambrian oceans? Some scientists have argued that environmental changes, such as increased oxygen levels, higher concentrations of key nutrients in the oceans and the development of predation, played important roles in driving the rapid diversification of life in this period.

The dramatic explosion of lifeforms at the beginning of the Cambrian raises fundamental questions about the nature of evolution that perplexed Charles Darwin. Even if the precise timescale involved was unclear in Darwin's day, the abrupt appearance of the Cambrian animals seemed real. How could the history of life take such a rapid leap from single-celled organisms to spectacularly diverse marine ecosystems swarming with swimming, crawling and burrowing animals? Darwin's grand theory of species origins proposed steady and ordered change. It did not accommodate such a spectacular evolutionary gear-shift. This became known as Darwin's dilemma.

The apparent abruptness of the transition to complex life troubled Darwin because it ran counter to all his rational instincts and to the huge body of evidence he had gathered to formulate his theory of evolution. He worried that it might be hijacked by Creationists, but later work has shown the fossil record to be much more complete than he thought, with abundant evidence for the work of evolution.

For example, we now know that after more than 2 billion years of prokaryotes (single-celled organisms with free-floating DNA that lack a nucleus), eukaryotes (organisms with membrane-bound nuclei, including all animals, fungi, protists and plants) appear in the rock record at least 2 billion years ago. Early eukaryotes were single-celled creatures, with greater complexity and multicellularity emerging around 1 billion years ago. Rocks in Arctic Canada dated to just over 1 billion years old contain fossils of an algae called *Bangiomorpha pubescens* that appears to be closely related to modern red algae. This is the oldest confirmed multicellular eukaryote; these fossils represent the oldest known organism that could reproduce sexually and the oldest multicellular organism that used photosynthesis. This was a major innovation in the evolution of life and major step on the way to the emergence of more complex animals.

One of the most famous sites that shows the wonders of the Cambrian explosion in extraordinary detail is high up in the Canadian Rockies. The Burgess Shale beds form one of the most important palaeontological sites in the world and were discovered in 1909 by Charles Doolittle Walcott (1859–1927), who was secretary of the Smithsonian Institution. Over many years of fieldwork Walcott and his collaborators, including his wife and children, collected some 65,000 specimens from the Burgess Shale. They form part of the fossil collections at the Smithsonian's National Museum of Natural History in Washington DC. These fossils preserve a remarkably detailed insight into how the animal kingdom was rapidly diversifying in the Cambrian Period.

The Burgess Shale fauna includes some bizarre forms with anatomical features that bear little resemblance to living animals.

In 1912 Walcott described one of the strangest and most iconic creatures from the Cambrian explosion. *Opabinia regalis* is an extinct shrimp-sized, soft-bodied arthropod with five stalked eyes that provided a wide field of vision in murky Cambrian seas; it wielded a grabbing clawed appendage that could pass food to the mouth. *Anomalocaris* (meaning 'strange shrimp') first appeared about 521 million years ago; it was a huge beast for its time with a hard-shelled, segmented body up to a metre in length, large compound eyes and two spiny front appendages to grab prey. While its fossils are widespread and often beautifully preserved, its bizarre body shape confused palaeontologists for decades. It was the apex predator of its day in the Cambrian ocean and one of the first known predators in Earth history. The Burgess Shale has preserved a bewildering range of hard- and soft-bodied animals in extraordinary detail, allowing us to peer into a wonderfully diverse marine ecosystem from half a billion years ago. Research on these Cambrian rocks and fossils continues apace today. The oldest known species of jellyfish was recently discovered in the middle Cambrian of the Burgess Shale.

For the first half of the twentieth century, palaeontologists were of the view that complex life began in the Cambrian Period. This provided a neat division between the Cambrian and the Precambrian. The latter had yielded only simple lifeforms like cyanobacteria and research into Precambrian life was commonly viewed as a palaeontological dead end. Why devote one's career to searching for microbes when there was much more excitement to be had studying the weird and wonderful creatures of the Cambrian oceans or the flesh-ripping dinosaurs of the Jurassic and Cretaceous? But where were the ancestors of the Cambrian fauna? If Darwin's theory of evolution was broadly correct, at least the later part of the Precambrian should also contain fossils of complex animals. But in Darwin's day these rocks were silent – they had not been sampled or studied in detail. Darwin's dilemma would not be solved for another hundred years.

Some palaeontologists have argued that the Cambrian fauna stands out because hard exoskeletons are much more likely to be preserved in the rock record. Thousands of species of trilobite have been described. The issue of fossil preservation bias is a particular challenge in the Precambrian before hard body parts had evolved. While it is true that soft-bodied creatures are much less likely to be preserved, this argument does not fully explain the Cambrian fossil record. There was certainly an explosion in hard-bodied animals in the Cambrian and most palaeontologists agree that the Cambrian 'explosion' is a real event. But we now know that it was an explosion with a long fuse.

In 1946, while exploring the Ediacara Hills in South Australia, geologist Reg Sprigg (1919–1994) discovered fossilised impressions of soft-bodied multicellular creatures in sandstones that were older than the Cambrian. These creatures would reshape our understanding of the evolution of complex life. Sprigg's sensational discovery led to the formal designation, in 2004, of the Ediacaran Period (635 to 538.8 million years ago), the first new period added to the geological timescale in over a hundred years.

Much attention has shifted to finding rocks of the Ediacaran Period, which is sandwiched between the Cryogenian (when Snowball Earth glaciations took place) and the Cambrian. This period is of critical importance for the study of life on Earth because this is when we see the first animals in the geological record. After some 3.5 billion years of microbes, the animal kingdom was about to get started.

Understanding life in the Ediacaran Period is crucial for unravelling the evolutionary processes that led to the emergence of complex animals and the explosion of diversity we see in the Cambrian Period. There has been a surge of interest in the animals of the Ediacaran over the last few decades as new *Lagerstätten* have been discovered including, notably, in Newfoundland, China and the White Sea coast of Siberia.

The first complex animals appear in the fossil record in the Ediacaran Period between about 600 and 572 million years ago, just after the last Snowball Earth glaciation. But there is still much uncertainty about their deeper origins because there is molecular evidence that puts their emergence as far back as 800 million years ago. If this is the case, these creatures or their precursors must have survived at least two Snowball Earth periods – the most extreme ice age conditions in Earth history. But where are the Cryogenian fossils?

The fossil record of this period is frustratingly incomplete. The huge glaciers of the Snowball Earth periods would have advanced across continental shelves and scoured away the fossil record. Creatures may have found refuge in the deep ocean where warm fluids emerged from mid-ocean ridges. Some may have adapted to life beneath the sea ice; indeed, animal life is surprisingly diverse on the seabed beneath Antarctic ice shelves today – communities of sea anemones, sponges and crinoids are common. These extremophiles are adapted to life in darkness and very cold water. The extreme environmental stresses of the Cryogenian oceans may have forced animals to become more complex.

The widespread and extreme glaciations of the Cryogenian would have eroded huge amounts of the Earth's crust. Since glaciers pulverise rocks into fine-grained particles, this would have released large quantities of key elements like phosphorus into marine ecosystems, especially towards the end of glacial periods when meltwaters flooded into the oceans. It seems counterintuitive to argue that Snowball Earth events may have nurtured the evolution of complex life on Earth, but the release of key nutrients in large quantities alongside increasing oxygen levels may have created optimum conditions to turbocharge the development of complex life in the Ediacaran.

The Cambrian explosion was a remarkable episode in Earth history that saw the emergence of a breathtaking diversity of complex marine animals. Radiometric dating shows that this episode played out over some 20 to 25 million years – a blink of the eye across the great sweep of Earth history but, from a biological perspective, plenty of time for evolution to do its work. The extraordinary evolutionary innovations that took place in the Cambrian oceans created early versions of every animal body plan that exists today. And it is now clear that animal species were diversifying much earlier; the ancestors of many of these Cambrian animals appeared during the Ediacaran. Animal life began in the Precambrian. And, while the fossil record of the Cryogenian has clearly been compromised, it is an intriguing possibility that the fuse that ignited the Cambrian explosion may have been lit all the way back beneath the ice of Snowball Earth.

# CHAPTER 16

# From Life in the Ocean to Life on Land

The colonisation of the land by plants and animals was a fundamental shift in the history of life on Earth. While the oceans were teeming with new creatures in the Cambrian Period, the continents were a bleak and largely lifeless rocky wasteland with scattered communities of bacteria and algae. It is difficult to be precise about when plants and animals first inhabited dry land because this was very likely a start-stop process, and the fossil record is incomplete.

The oldest evidence we have for a creature walking on dry land comes from trace fossils discovered in a disused quarry in Kingston, Ontario. Exposures in late Cambrian sandstones reveal multiple criss-crossed trackways made by creatures similar to a large centipede. We can be sure this animal was on dry land some 490 million years ago because structures in the ancient host rocks were formed in a coastal dune setting. It is very unlikely that this animal was fully terrestrial, but these are the oldest known footprints on land.

Adapting to a land-based life posed many new challenges: plants and animals had to cope with exposure to strong sunlight and with rapid shifts in temperature and the availability of water. Poorly developed soils offered only a limited supply of nutrients. Outside a watery world, animals had to evolve new ways of seeing, hearing, breathing and moving. All this took tens of millions of years.

The Palaeozoic Era began with the Cambrian explosion and ended with a catastrophic mass extinction at the end of the Permian – the greatest loss of lifeforms the Earth has ever seen. The Palaeozoic was also a time of profound change in physical geography, encompassing large-scale glaciation, drifting continents, episodes of extensive volcanic activity and big fluctuations in sea level. By its close, the Palaeozoic Era had seen the establishment of a truly global biosphere, with the first complex ecosystems on land.

As biodiversity increased on the continents and landscapes became ever greener, vegetation trapped large amounts of carbon and began to have an increasingly important impact on the composition of the atmosphere. It was a bumpy ride punctuated by extinctions of various sizes, but during the course of the Palaeozoic Era we see the emergence of a fully functioning and dynamic Earth system with diverse land-based ecosystems playing a full part. For the first time in Earth history, plants could change the climate.

The Cambrian wasn't just about the emergence of new creatures. It also saw a series of extinction events culminating in a major extinction at its close that wiped out many of the species of trilobites and molluscs on shallow marine shelves. Sedimentary rocks from this period contain elevated levels of sulphur and carbon, which points to major falls in oxygen levels in seawater. These oxygen crashes have been linked to episodes of global warming and phases of intense volcanic activity. Many creatures in deeper water struggled to breathe and may have suffered poisoning by hydrogen sulphide: the latter smells of rotten eggs and is produced when organic matter decomposes in the absence of oxygen.

The extinction at the close of the Cambrian nevertheless provided evolutionary opportunities. The Ordovician Period began 486.9 million years ago and lasted for almost 44 million years. Intense greenhouse warming produced a very warm climate and high sea levels (because glaciers were scarce) during its early part.

As in the Cambrian, the Ordovician seas were rich with life. The Ordovician saw the emergence of highly diverse marine ecosystems in all parts of the oceans. Many new species flourished and expanded their geographic ranges. The first coral reefs appeared – these are some of the richest habitats in today's oceans. The Ordovician also saw a remarkable diversification in ways of living and moving, with floaters, swimmers, filter feeders, burrowers, walkers and scavengers. New communities and food webs emerged at all depths of the ocean.

This burst of life has been called the Great Ordovician Biodiversification Event (GOBE). It saw a big jump in the diversity of tiny creatures collectively known as *phytoplankton*. Their habitat is predominantly the upper parts of the ocean where they can use sunlight to generate energy via photosynthesis. Their success provided a rich food supply at the base of the marine food chain as ocean productivity boomed. Trilobites were the dominant scavengers on the Ordovician sea floor, while huge nautiloids (a distant relative of the modern squid and octopus), with their long, straight chambered shells, became the top predators in open water. Some sea creatures took on gigantic proportions. The nautiloid *Endoceras giganteum* from the Upper Ordovician of North America had a body length of 9 metres. They may have been ambush predators that prowled the seabed. The oceans had never been more diverse.

The first graptolites (from the Greek *graptos*, meaning 'written', and *lithos*, meaning 'rock') were slender creatures attached to rocks on the sea floor or rooted in the mud. Their elongate fossils often resemble pencil sketches. In the early part of the Ordovician these simple stick-like animals became free floating and fed on phytoplankton. Graptolites were among the first creatures to exploit the open sea habitat and they did this very successfully in huge, interconnected colonies. Billions of these creatures so dominated the upper layers of warm tropical oceans that their fossils are found on every continent apart from Antarctica.

During the Ordovician and Silurian, Earth's geography was dramatically different from today. The northern hemisphere was almost entirely ocean while a giant landmass called Gondwana straddled the South Pole, with several smaller landmasses closer to the equator. Sea level was much higher than today, with large areas of what is now dry land – including all of what is now North America – submerged under shallow seas. The hard limestone rocks that were much later uplifted to form many Himalayan summits were deposited in a warm tropical ocean during the Ordovician Period. Thick fossil-rich mudstones were deposited across huge areas – these rocks preserve evidence of a remarkable radiation in marine species.

The Ordovician is a key period for the development of life on Earth because the GOBE saw a spectacular increase in marine biodiversity across *all* taxonomic levels. This extraordinary radiation saw new species living in habitats at all depths across the vast expanse of the Earth's oceans.

As Earth gradually cooled during the Ordovician, oxygen levels increased in the ocean, supporting more complex food webs. The breakup of continents and volcanic activity may have increased the delivery of nutrients like phosphorus into the oceans, boosting productivity at the base of the food chain. The GOBE has even been linked to the disintegration of a huge asteroid around 470 million years ago in the Middle Ordovician that led to frequent impacts of kilometre-wide asteroids. Some researchers have argued controversially that, rather than the well-documented negative effects of large asteroid impacts, more minor and persistent impacts on Earth could actually promote biological diversity by creating new niches and habitats.

Current understanding of land-based plant communities during the Ordovician is rather sketchy because the fossil record is limited. There is some evidence from tiny fossil spores suggesting the first land plants evolved from green algae during the Lower to Middle Ordovician. By the end of the Ordovician Period, simple plants like mosses and liverworts had begun to establish communities on the margins of coastal wetlands.

The Silurian Period followed the Ordovician, beginning some 443.1 million years ago. It was a period of significant firsts in animal and plant evolution, including the origin of vascular plants with stems and leaves. The earliest fossils of the oldest known plant with a stem, *Cooksonia*, come from Silurian rocks in the Czech Republic. Picture a leafless Y-shaped stick, standing a few centimetres tall. *Cooksonia* had primitive vascular tissue that could transport water and nutrients, helping it survive on land. This innovation heralded a huge explosion in terrestrial plant life with a tenfold increase in the number of plant species. The earliest life on land resembled today's moss- and lichen-rich ecosystems, with simple, small-stature plants and animal life dominated by arthropods. These included primitive millipedes and centipedes, and a group of animals called trigonotarbids – similar in appearance to spiders with segmented bodies and eight legs but lacking the ability to spin webs. Larger plants leading to forest ecosystems on a scale that we would be familiar with today began to emerge by the Middle Devonian.

The Devonian Period (419.6 to 358.9 million years ago) was a key period for the evolution of land-based ecosystems. It has even been said that the Devonian is to botanists what the Cambrian is to zoologists because it is when land plants really took off, rapidly diversifying, expanding their range, and increasing in size and complexity with well-developed root systems. Vast areas of the continents became green during the Devonian as many new kinds of plants evolved and the first forests emerged with trees reaching heights of several metres.

Once plants became established on land they transformed the Earth's surface. Early plants like mosses and liverworts developed simple root-like structures that secured them in the ground. Various species of fungi emerged that helped to create soils with organic litter that could store water in the ground. This may have helped streams to flow year-round. Fungi have played a key role as decomposers since the first terrestrial ecosystems emerged. They also form mutually beneficial symbiotic relationships with plants, colonising their roots and helping them absorb nutrients from the soil, while being provided with sugars made by the plant via photosynthesis. This relationship may have been crucial to enabling plants to colonise and evolve on land, including the evolution and spread of vascular plants, which was a major innovation of this period. The Rhynie chert in Scotland provides a 410-million-year-old snapshot of an ancient ecosystem from the Lower Devonian period, with some of the oldest and best-preserved land plants. It is a remarkable site where geysers from volcanic hot springs sprayed silica over plants and animals, preserving them in stunning detail before they decayed. It shows early ecosystems with plant–fungi symbiosis, primitive root systems and soil development.

This greening of the land modified Earth's water cycle, promoted soil development and extracted huge amounts of carbon dioxide from the atmosphere. When plant material is buried in the ground and does not decompose it changes the balance of the carbon cycle so that the concentration of oxygen increases in the atmosphere at the expense of carbon dioxide. As the free oxygen content of the atmosphere increased, this helped animals to survive out of water.

The oldest known air-breathing land animal is a tiny millipede, *Pneumodesmus newmani*. It was discovered by (and named after) Mike Newman, a bus driver and fossil enthusiast from Aberdeen, in sedimentary rocks on the coast of northeast Scotland. It dates to the Silurian Period some 428 million years ago. The fossil is only a centimetre across, but tiny holes called spiracles are well preserved, showing that this animal breathed oxygen from the air. This was a creature that *lived* on the land – it would not have been able to breathe in the sea because the holes would have filled with water. Millipedes are important ecosystem engineers because they shred plant material, making it available for further breakdown by microbes. These processes release nutrients such as calcium and improve the quality of the soil. This was essential groundwork for the later emergence of larger and more complex plants with deeper root systems.

The Ordovician is a fascinating period for the study of evolution because one of the largest and fastest biodiversification episodes is followed by a mass extinction when complex life was almost wiped out. The end of the Ordovician saw the first mass extinction of marine animals when about 80 per cent of species living in the warm shallow seas surrounding the continents were lost. This was the second largest mass extinction in Earth history. There is good evidence pointing to two closely spaced extinction pulses.

Several theories have been put forward to account for this great dying, including a large-scale glaciation of the southern Gondwana landmass, not unlike the glaciation of Antarctica today. As the Ordovician ice sheets thickened, and more and more water was stored frozen on the land, global sea levels dropped, leaving the highly productive shallow sea ecosystems of the continental shelves high and dry. The cause of the glaciation has been linked to the expansion of land plants locking up carbon dioxide from the atmosphere and weakening the greenhouse effect. As the climate cooled, the polar south crossed a threshold for the onset of large-scale glaciation.

The emergence of complex, land-based life was undoubtedly one of the major advances in the history of life on Earth. Evolutionary biologists call this process *terrestrialisation*. We don't know all the details, but it was a slow and geographically uneven process with stops and starts that took many millions of years. As land-dwelling plants and animals began to gain a foothold during the Ordovician and Silurian, it was mainly in coastal wetland habitats and along the margins of rivers. But not everywhere was turning green. The landscapes of the dry continental interiors remained essentially lifeless for the first half of the Palaeozoic.

# CHAPTER 17

# The Global Greenhouse
# and Ancient Climates

When we hear the greenhouse effect being discussed on the news, it's mostly gloomy reports of its role in the rapid climate warming of recent decades and the dire consequences for our planet if it is not curbed. But Earth's *natural* greenhouse effect is in fact a crucial life-support mechanism. It plays a critical role in keeping our world habitable. Without the heat-trapping properties of greenhouse gases like carbon dioxide, our planet would be stuck in a permanent ice age and complex life may never have emerged. Earth history would be rather dull!

We have natural greenhouse warming to thank for raising today's average global temperature by a massive 35 degrees, to just above 15°C. Without the insulating effect of greenhouse gases, this figure would be −20°C – a freezing-cold Arctic scenario, roughly the average annual temperature today on the southern part of the Greenland ice sheet. Greenhouse gases are not only essential for human survival and for the existence of most lifeforms, they have also played a key role in regulating Earth's temperature for billions of years.

Carbon dioxide ($CO_2$) and water vapour ($H_2O$) are the most important greenhouse gases in today's atmosphere, alongside nitrous oxide ($N_2O$), methane ($CH_4$) and ozone ($O_3$). To see what makes these gases special and how they create the global greenhouse, we need to look first at their impact on Earth's energy balance.

The solar energy received by the Earth is *short-wave* energy in the visible and ultraviolet part of the electromagnetic spectrum. This is why you apply cream with an ultraviolet filter to protect your skin from the harmful effects of the Sun's rays. About one-third of this incoming short-wave energy is reflected from the Earth's upper atmosphere back into space and plays no part in warming the planet. The rest passes through the atmosphere and warms the land surface and oceans. These warm surfaces emit *long-wave*, or infrared, radiation back into the atmosphere. This radiant heat can be viewed with an infrared camera. If you hold your hand close to a black tarmac road on a sweltering day you will feel the heat radiating from that surface. This outgoing longer-wavelength heat does not simply radiate back into space, it is trapped in the Earth's atmosphere by greenhouse gases.

The gases that make up Earth's atmosphere have very different properties. Oxygen ($O_2$) and nitrogen (N), for example, which together make up about 99 per cent of the atmosphere, allow infrared waves to pass freely. They do not provide any insulation. In contrast, molecules of carbon dioxide ($CO_2$), methane ($CH_4$) and water ($H_2O$) are made up of more than one element and three or more atoms. Because they have more complex molecular structures that can stretch and bend, they are able to interfere with and absorb a much wider range of energy wavelengths. Even though they are present in tiny concentrations, these gases regulate Earth's energy balance.

In the middle decades of the nineteenth century, two scientists working independently on opposite sides of the Atlantic Ocean made fundamental contributions that led to the discovery of the greenhouse effect. Eunice Newton Foote (1819–1888) carried out elegantly simple experiments with gases and heat in the 1850s and was the first to recognise that changes in the concentration of carbon dioxide and water vapour in the atmosphere could change its temperature. One experiment showed that a glass cylinder filled with $CO_2$ and placed in direct sunlight trapped more heat and stayed warmer for longer than a cylinder filled with normal air. Foote realised that air with a higher proportion of carbon dioxide would result in an increased temperature. She had discovered the key mechanism behind human-made, or anthropogenic, global warming.

Her findings were presented in 1856 at the tenth meeting of the American Association for the Advancement of Science in Albany, New York. Foote's discovery provided the theoretical basis for what we now call the greenhouse effect. At around the same time, the Irish physicist John Tyndall (1820–1893) conducted similar experiments that showed that certain gases could trap heat within the atmosphere. Tyndall provided a more detailed physical explanation for the observed warming, recognising it was long-wave radiation that was absorbed in the atmosphere. Like Foote, he saw the significance of his findings for Earth's climate when he remarked that these gases 'would produce great effects on the terrestrial rays and produce corresponding changes of climate'. For many decades, Tyndall's name was exclusively associated with the discovery of the greenhouse effect because the full significance of Foote's pioneering work was only recognised in the twenty-first century.

Let's focus on carbon dioxide, an odourless, tasteless and colourless gas. This gas traps a portion of the emitted long-wave radiation because molecules of $CO_2$ absorb energy at wavelengths that overlap with those of infrared energy. Like the other greenhouse gases, it acts as a blanket that limits the escape of heat back into space. As the concentration of carbon dioxide in the atmosphere increases, the blanket becomes more effective at trapping heat so, as Foote and Tyndall predicted, the planet gets warmer. Conversely, if the concentration of $CO_2$ decreases, the greenhouse effect is weakened, more heat escapes to space and the planet cools. We are living in an era of greenhouse warming because huge volumes of $CO_2$ have been pumped into the atmosphere by the burning of fossil fuels since the onset of the Industrial Revolution (Chapter 37). Here, let's explore the deeper geological history of climate changes driven by shifts in the strength of the greenhouse effect.

There are many examples in nature where trace amounts of a substance can exert a major influence on a system. Even though carbon dioxide accounts for only 0.04 per cent of the Earth's atmosphere (and rising), its heat-trapping properties have an enormous impact on the average global temperature. Methane accounts for even less. It is only 0.00017 per cent of the atmosphere by volume, but its presence is important because it is roughly twenty-eight times more effective than carbon dioxide at trapping heat. It is thought that methane played a key role in keeping the young Earth warm during the Hadean and Archean when the Sun was weaker, and it made up a much greater proportion of the primitive atmosphere. Methane has also played a role in some of the intense greenhouse warming episodes that have punctuated the last 500 million years of Earth history (Chapter 25).

The rock and fossil records show that global climate has regularly shifted from warm to cool and back again. Greenhouse gases play a key role in regulating our planet's energy balance and average temperature. They have done this throughout Earth history and changes in the concentration of carbon dioxide have been especially important. When Eunice Foote conducted her experiments in the 1850s, the carbon dioxide concentration in the atmosphere was about 280 parts per million (ppm). It passed 425 ppm in 2024 (and continues to rise). The last time atmospheric $CO_2$ was above 1,000 ppm was in the middle of the Eocene Epoch about 45 million years ago, when the average global temperature was about 13°C warmer than today. This is a super-greenhouse scenario. Earth was largely ice-free at this time and sea levels were 70 metres higher than today. Conversely, computer modelling of the Neoproterozoic climate suggests that the onset of Snowball Earth episodes needed carbon dioxide concentrations below 100 ppm. Snowball Earth conditions then flipped to extreme hothouse warming. Earth scientists know that there are many factors that influence the climate, but a good deal of attention has been focused on explaining such large changes in atmospheric carbon dioxide over extended geological timescales.

While climate extremes get most of the attention, a key feature of Earth's long-term climate history is its remarkable propensity to avoid getting permanently stuck in one of these hot or cold extremes. Some geologists have argued that Earth has a thermostat set to 'habitable' to prevent it getting trapped in an icehouse or hothouse episode. In other words there are feedbacks in the Earth system that bring the planet back to more typical 'average' conditions of not too hot and not too cold. To explore this Goldilocks idea we need to think about plate tectonic processes and the role they play in the long-term carbon cycle.

There is a good deal of evidence to suggest that, over very long timescales (tens and hundreds of millions of years), plate tectonic processes have functioned as a global thermostat that prevents the Earth from getting stuck in an extreme climate state. Research has shown that feedback mechanisms ultimately driven by plate tectonic processes can put the brakes on both extreme warming and extreme cooling. They have kept Earth habitable by regulating the carbon cycle. This is sometimes called the slow carbon cycle.

Various gases, including carbon dioxide and water vapour, are added to the atmosphere by volcanoes. Such volcanic degassing is a process that takes place continuously, even when volcanoes are not erupting. It is the main way that carbon is transferred from the mantle to the atmosphere over geological timescales. The amount of volcanic activity taking place over time is effectively controlled by plate tectonic processes. When the movement of the tectonic plates speeds up, the amount of volcanic activity increases and more carbon dioxide is pumped into the atmosphere. This strengthens the greenhouse effect and Earth gets warmer. This process has triggered some of the big extinctions in the fossil record (Chapter 20).

Since volcanic activity has been in progress for most of Earth history, one might expect the carbon dioxide content of the atmosphere to steadily increase over time and the planet to get warmer and warmer. This has happened on Venus, where the atmospheric carbon dioxide concentration is 97 per cent! The geological record tells us that this is not the case on Earth. Carbon dioxide is removed from the atmosphere in two main ways.

Carbon is cycled through the Earth system via a series of sources, pathways and sinks, including the rock record, the atmosphere, the oceans and the biosphere. Carbon dioxide is extracted from the atmosphere by plants during photosynthesis. When plants convert sunlight into energy they transform $CO_2$ and water into sugars and oxygen. Plants store this carbon until they die. In the Carboniferous Period (358.9 to 298.9 million years ago) huge areas of land were covered by swamps and forests. Dead plant material didn't fully decompose in the low-oxygen swamps. Instead, it was buried and preserved in peat layers, which over time hardened into coal. Over millions of years this process locked away (sequestered) carbon that would otherwise have been returned to the atmosphere, so global climate cooled as atmospheric $CO_2$ concentrations fell. This is often called the biological pump, taking carbon dioxide out of the atmosphere.

The other main mechanism is driven by weathering and erosion. A warmer climate tends to generate higher rainfall because a warmer atmosphere holds more water. This increases chemical weathering rates so that more rock material is broken down and transported to the oceans. Rainfall is a mild acid because it mixes with carbon dioxide in the atmosphere. This acid attacks rock minerals and the soup of weathering products is ultimately transported by rivers to the ocean, where the carbon is used to build shells and corals and eventually carbonate rocks. It can take a long time, but this process will lead to global cooling when carbon is taken out of the atmosphere and stored in carbonate rocks on the seabed at a faster rate than it is added by volcanic activity.

When tectonic plates clash and build large mountain ranges, the climate can cool more rapidly because weathering and erosion take place at much higher rates. The creation of large mountain ranges will also increase the extent of snow cover, and this will reflect some solar energy back into space. As mountains go up, global temperature goes down (Chapter 26).

So processes of erosion and weathering function as a thermostat. When the Earth heads into cold-climate mode with large-scale glaciation, chemical weathering and erosion tend to fall and further cooling is halted. This is a simplification of a complex set of processes operating at a grand scale, but this internal regulation of the carbon cycle is a remarkable feature of Earth's shifting climate over geological timescales.

The balance between the energy received from the Sun and energy loss back into space has always been a vital control on the temperature of our planet. As Foote's and Tyndall's experiments showed in the nineteenth century, the concentration of carbon dioxide in the atmosphere plays a key role in regulating Earth's temperature. Over very long timescales feedbacks in the carbon cycle have prevented global temperatures from getting stuck in extremes of hot or cold that would pose a threat to life on Earth. But these natural processes do not happen quickly enough to help humanity tackle the rapidly warming climate of the modern era.

# CHAPTER 18

# Fishes, Four Limbs and Forests

A few days before Christmas 1938 the curator of the East London Museum in South Africa took a phone call from the captain of a fishing boat who had hauled up the most extraordinary creature. Marjorie Courtenay-Latimer (1907–2004) stopped what she was doing and jumped in a taxi to the harbour. She later recalled the encounter that would make her famous: 'I picked away at the layers of slime to reveal the most beautiful fish I had ever seen ... it had an iridescent silver-blue-green sheen all over. It was covered in hard scales, and it had four limb-like fins and a strange puppy-dog tail.' The fish was a whopper, about 5 feet long. The taxi driver begrudgingly agreed to take the huge slimy beast back to the museum.

Courtenay-Latimer had never seen a fish like this one and it wasn't in any of her books. She made a detailed sketch and sent it to a chemistry professor friend at Rhodes University who had a long-standing fascination with fish. As James Smith (1897–1968) studied the drawing he couldn't believe his eyes. He soon realised the fish was a coelacanth (pronounced *see-la-kanth*), a group of fish that first appeared in the Devonian Period some 400 million years ago. This weird-looking fish was thought to have vanished 66 million years ago in the catastrophic mass extinction that wiped out the dinosaurs and marine reptiles (Chapter 23). At least that's what the palaeontology textbooks said. Courtenay-Latimer's discovery caused a sensation – this primitive beast was very much alive and swimming in the Indian Ocean. For the natural sciences, it was one of the most important discoveries of the twentieth century.

There are more than 175 fossil species of coelacanths known from the Palaeozoic and Mesozoic, with the highest number of species in the Triassic Period. The Devonian coelacanths inhabited warm, shallow marine habitats with reefs and lagoons; some may have tolerated brackish conditions in estuaries. We know of two living species of coelacanths, and it is not surprising they escaped detection for so long. Today they live in the murky deep waters close to tropical volcanic islands where divers have observed them grouped inside caves formed in solidified lava. Scientists get so excited about the living coelacanths because they help us to understand the process of evolution. Coelacanths have been called living fossils because they have retained anatomical features that are not present in any other living fishes. A hinge in the skull, for example, allows it to enlarge its mouth opening. No living vertebrate has this feature. Even more intriguingly, the coelacanth has muscular limb-like fins (with bones inside) that paddle in a synchronised fashion like the legs of some four-legged animals.

The Devonian Period (419.6–358.9 million years ago) followed the Silurian and lasted for some 60 million years. It is often called the Age of Fishes because important new groups of fish emerged, and the fossil record of fish is especially rich. Land-based ecosystems also became much more complex as plants and insects diversified and the first seed-bearing plants evolved. This period is especially notable for the emergence of the first tetrapods (land-living four-legged verte-brates), the first terrestrial arthropods and the first forests.

Coral reefs also became widespread, providing diverse ecosystems in the warm shallow waters that fringed the continents.

Armoured fish called placoderms – with hard plates like medieval helmets covering their heads – came to dominate the open ocean. Some grew to an enormous size and became fearsome apex predators. Many placoderms lived in shallow seas too, especially along continental shelves and reefs where food supplies were abundant. Some species adapted to freshwater habitats, living in rivers, lakes and swampy floodplains.

When researchers began unearthing the massive, armoured fossil skulls of the placoderm *Dunkleosteus terrelli* in the nineteenth century, they assumed it had a long shark-like body and estimated its body length to be about 9 metres. Because the hard-plated skulls are often well preserved, while the rest of the body is not, palaeontologists had to guess what its body looked like. The latest research has shown that this fish was not long and streamlined, it was short and super chunky – more tank than stretch limo. Nonetheless, *Dunkleosteus terrelli* was a hugely successful predator in the Devonian seas and very probably the biggest animal that had existed up to that time. Placoderms were among the first jawed fish but did not develop teeth. Weighing between one and three tonnes, *Dunkleosteus terrelli* was the first super predator in the oceans, with ferocious scissor-like jaws that were twice the size of those of a great white shark. Instead of teeth they had razor-sharp bony plates that snapped shut like a mantrap with the most powerful bite force of any fish.

Placoderms are among the oldest animals whose colour we have information about. Fossils of a river-dwelling species discovered in Devonian rocks in Antarctica included preserved pigment cells that revealed a red back and silver underside. This suggests Devonian fish could see colours. Placoderm fossils have been found in river and lake sediments – they were some of the earliest vertebrates to adapt to life in fresh water.

Lobe-finned fishes and the armoured placoderms dominated the Devonian ocean while ray-finned fishes and sharks were in the minority. The end of the Devonian is marked by a mass extinction that radically reorganised life in the oceans. This was a prolonged and complex extinction rather than a single catastrophe, with multiple potential causes including global cooling and sea-level fall as well as volcanic activity changing the composition of the atmosphere and ocean chemistry. The spread of forests and soil development (Chapter 16) may have flushed nutrients into the oceans, which increased productivity and reduced the supply of oxygen. Placoderms disappear from the fossil record and lobe-finned fishes were largely replaced by ray-finned fishes, which account for 99 per cent of today's fish species, from clownfish to sea bass. They have bony skeletons and fins supported by bony spines that can be stretched like a fan.

Every now and then a fossil is discovered that is difficult to classify because it displays a mix of characteristics that place it between two groups of animals. These are known as *transitional* fossils. In the summer of 2004, the beautifully preserved fossil skeleton of a new species of ancient fish was discovered in pink Devonian sandstones on Ellesmere Island in the tundra of Arctic Canada. About 375 million years ago, northern Canada lay close to the equator, so would have had a tropical climate. The creature was named Tiktaalik, which, in the Inuktitut language, means 'large freshwater fish'. But it wasn't just another fossil fish from the Devonian – careful study of its anatomy revealed that it was a mix of fish and tetrapod.

Tiktaalik was a big beast with a body length of some 2.75 metres. It was a predator with a curious mix of features: a crocodile-like head and sharp teeth, a gill slit, lungs, scales and, like the coelacanths, lobe-like fins. But it also had a robust pelvis and sturdy limb bones that could have been used for swimming in deeper water but would also have allowed it to waddle around in shallow water on riverbeds. With a sturdy ribcage, Tiktaalik had lost the bone structure seen in fish that helps breathing with gills. It had a neck which suggested it could hold itself up in shallow water to look above the surface. This creature literally took the first steps that led to four-legged land animals.

Tiktaalik is a compelling demonstration of evolution in action. Its anatomy represents an intermediate stage when fish had developed most of the anatomical features needed to become four-legged animals. Its skeleton provided new insights into the evolutionary process of how fleshy fins became limbs, and it had acquired the ability to breathe air – a key step along the path to life on the land. It is one of the most important fossil animals ever discovered because it plugged a critical evolutionary gap – between fish and the first land-living tetrapods – that palaeontologists had been seeking to fill for decades. Tiktaalik shows us that the first tetrapods were still aquatic creatures that relied on water for breathing, reproduction (laying eggs in water) and most move-ment, because they were better swimmers than walkers. The leap to the land was a gradual evolutionary process.

The fish–tetrapod transition was a critical milestone in the history of life on Earth. Tiktaalik's anatomy represents a crucial stage in the evolution of all limbed vertebrates, including humans. And the detailed study of Devonian fossils has shown how certain features of human anatomy can actually be traced back to features found in ancient fishes. For example, all mammals have three tiny bones in the middle part of the ear that help transfer sounds from the eardrum to the inner ear and brain. One of these bones, the stapes, is the smallest bone in the human body, and had a rather different role 380 million years ago: it was part of the jaw anatomy of Devonian fish. Over tens of millions of years these bones became smaller and changed function as animals that left the water developed the ability to hear sounds in air.

We can never know why some animals left the ocean for a life on the land. Perhaps they were escaping from predators or exploiting the availability of new sources of food. As well as the classical fossil evidence, there are important sites with well-preserved tetrapod trackways that can tell us about their movement and habitat. On the rocky shores of Valentia Island off the Atlantic coast of south-west Ireland there are several beautifully preserved trackways from the Middle Devonian showing that populations of four-legged animals walked along the muddy margins of rivers in a warm, semi-arid climate.

Devonian rocks in the sea cliffs of southwest England that date to 390 million years ago contain fossils of the oldest known forest.

This forest was dominated by a tree that looked like a palm but with a crown of long, woody twigs. Leaves did not become widespread until the very end of the Devonian some 30 million years later. As the amount of plant matter increased and oxygen levels in the atmosphere continued to rise, the Devonian had all the ingredients for fire. While the fossil record shows some localised evidence for burning in the Silurian, the coastal forests of the Upper Devonian record the first wildfires in Earth history.

As plants increased in size above and below ground in the Devonian they became a major new carbon store. As they pulled more carbon dioxide out of the atmosphere they changed its composition. This process continued into the Carboniferous Period (358.9 to 298.9 million years ago), when forests expanded and trees grew even larger. As the concentration of oxygen increased, the atmosphere could support larger and more complex animals on land – it was only in this period that tetrapods became fully adapted to life out of water. This atmospheric process involved important feedbacks, with the increase in oxygen accelerating the pace of plant and animal evolution. This is the Earth system in action. The spread of forests in the Devonian and Carboniferous was a key stage in Earth history. Today land plants lock up vast amounts of carbon and help to modulate the climate; they account for about 80 per cent of all the biomass (the weight of all living things) on Earth.

The Carboniferous forests not only had giant trees, they also hosted millipedes the size of crocodiles and dragonflies the size of magpies. The emergence of these giant insects has been explained by the presence of an oxygen-rich atmosphere. Insects do not breathe like mammals; they do not have lungs. They take in air through a network of tiny tubes called tracheae via tiny holes in the walls of the body called spiracles. When the concentration of oxygen in air increases, they can get more oxygen into their bodies because the distance oxygen can travel down these tubes increases. This is a limit on insect body size. The Carboniferous atmosphere was about 30 per cent oxygen compared with about 21 per cent today. Because big bugs have longer tracheae, higher concentrations of oxygen are needed to reach all parts of the body.

Carboniferous means 'carbon-bearing', reflecting the thick deposits of coal that built up during this period. Huge areas of the continents were covered by swampy wetlands where giant trees and other plants were buried and compacted into peat and eventually hard shiny black coal. Most of the world's coal formed during the later part of the Carboniferous. We are now rapidly warming our world by burning the carbon from giant trees that cooled Earth's climate over 300 million years ago.

# CHAPTER 19

# Volcanoes Past and Present

When Roman ships sailed past the tiny western Mediterranean island of Vulcano, their crews often saw fountains of glowing magma. From time to time they witnessed deafening explosions and clouds of ash. The Romans believed the island was the chimney of a subterranean workshop where Vulcan, god of fire, forged swords, shields and thunderbolts; the loud blasts and molten rocks told them he was labouring deep inside the Earth. Vulcano is part of a cluster of volcanic islands to the north of Sicily that have been active for thousands of years. The most recent explosive eruption, in 1888–90, saw rhyolite blocks cannoned hundreds of metres into the air. This island is the origin of the word volcano.

Ever since the formation of our planet, volcanic processes have played a major role in the evolution of landscapes and climate; they release huge amounts of energy and matter from Earth's hot interior and supply water vapour, carbon dioxide and other gases to the atmosphere. They destroy and rebuild landscapes. Volcanic processes have created about 80 per cent of the Earth's crust. While

they have helped shape and sustain life, volcanoes have also triggered mass extinctions and, in the age of humans, represent one of
Earth's most deadly natural hazards.

Volcanology is the study of volcanoes. Molten rock below
ground is called magma but when it breaks through to the surface,
either on land or below the ocean, it is called lava – or ash if it
comes out as particles of fine rock and glass. A volcano is an accumulation of the materials that have erupted (explosively or
passively) from a vent or fissure at the Earth's surface. Volcanoes
are both majestic and violent. Whether active, dormant or extinct,
they have inspired wonder and terror for millennia.

According to the United States Geological Survey (USGS), there
are about 1,350 active volcanoes on Earth. Around five hundred of
these have erupted since written records began and, on any given
day, somewhere in the world about twenty volcanoes are actively
erupting. It is difficult to count Earth's volcanoes because many lie
deep on the ocean floor. Some of these submarine volcanoes have
been observed from submersible vehicles, but most have never
been seen by humans. A recent study discovered 135 volcanoes
under the West Antarctic ice sheet buried beneath over 3 kilometres of glacial ice. Scientists are concerned that the heat generated from volcanic processes is melting the base of the ice sheet
and contributing to its instability.

Plate tectonic processes control the location of most volcanoes,
with about 75 per cent of Earth's active volcanoes found along
the Pacific Ring of Fire. Most explosive volcanoes are associated
with subduction zones where oceanic plate sinks into the mantle.
As the plate sinks it releases water that provokes melting of the
surrounding mantle rock. Magma rises up through the crust and
some of it makes it all the way to the surface. Tectonic setting is a
key control on eruption style because it influences how readily
magma flows and how much gas it contains.

Deep-seated hot plumes of the mantle can lift up plates, causing
them to rupture. This process is called rifting and can lead to the
creation of a new plate boundary peppered with volcanoes. The best
example of this process is the East African Rift system where the

African continent is splitting. Ethiopia sits within this great rift and has sixty-five volcanoes. About half of its rapidly growing population of some 129 million people live within 100 kilometres of a volcano.

Volcanoes come in various shapes and sizes, but there are two main kinds. A *stratovolcano* is made up of successive layers of ash, cinders and solidified lava. Sometimes called composite volcanoes, they are typically steep-sided, cone-shaped and symmetrical, with a characteristic summit crater that contains a central vent. Mount Fuji, Mount St Helens and Mount Vesuvius are classic examples – majestic, stand-alone peaks that dominate the landscape. Stratovolcanoes can reach great heights and many have year-round snow-capped summits and even glaciers. The highest are found in the Andes.

The second major type is a *shield volcano*. These tend to have milder eruptions with runny lava that flows easily over gentle slopes. The flanks of these volcanoes are long relative to the summit height – like a round shield on its side. Shield volcanoes are the largest volcanic landforms on Earth and the most famous is Mauna Loa on Hawaii. Its summit crater stands more than 4,170 metres above the Pacific Ocean, while the full height of the volcano down to the ocean floor exceeds 9,100 metres. That makes it about 250 metres higher than Mount Everest! Mauna Loa has erupted thirty-four times since the middle of the nineteenth century. Many of these eruptions produced large volumes of lava that flowed for long distances. The Hawaiian volcanoes are distinctive because they are located in the middle of the Pacific plate, about as far away from a plate boundary as it is possible to be. They have formed over a deep-seated 'hotspot' where the mantle is unusually hot.

Volcanoes erupt violently if pressure builds up when gases in the magma cannot escape. Stratovolcanoes are most likely to erupt explosively because they produce viscous, slow-moving, vent-clogging magma. They also tend to be located in subduction zones, where magma contains lots of dissolved water that can fizz out on eruption. The resulting explosive eruptions produce large quantities of ash and pumice. I remember watching the spectacular eruption of Mount St Helens in Washington State, USA, on television news when I got home from school in 1980. Within just ten minutes the ash column reached a height of 20 kilometres. Two weeks later, the ash cloud had circled the globe.

Krakatau is a small volcanic island that sits on part of the subduction zone where the Indo-Australian plate sinks below the Eurasian plate. It is one of more than 130 volcanoes in Indonesia. The Krakatau eruption in August 1883 is one of the deadliest on record, with more than 36,000 victims. The eruption was about 10,000 times more energetic than the atomic bomb that devastated Hiroshima; its explosive blast was the loudest sound in recorded history, heard even in Australia, over 4,500 kilometres away. It also generated the most destructive volcanic tsunamis ever recorded. Hundreds of villages in western Java and eastern Sumatra were deluged and waves over 30 metres high washed ships 3 kilometres inland. Blips in sea level were recorded around the world, including at tide gauges in England, Alaska and San Francisco.

After the Krakatau eruption, deep-red twilights were observed around the world. They were produced by exceptionally fine sulphurous dust in the atmosphere. The wavy red-orange sky painted a decade later by Edvard Munch in his most famous work *The Scream* (1893) may have been inspired by this eruption.

It is exceedingly difficult to get precise figures on casualties from catastrophic volcanic eruptions. It has been estimated that some 280,000 people have been killed by volcanoes since the year 1500. The USGS lists just six eruptions accounting for 217,000 deaths in that period. The hazard from volcanoes is real – today about 60 million people live within 10 kilometres of a volcano and 1 billion within 100 kilometres. One of the ironies of volcanoes is that they produce very fertile soils that lure people to live around them. The vineyards on the slopes of Mount Etna produce some of the finest wines in Italy.

Managing the risk posed by volcanoes is hugely challenging because long periods of inactivity can be broken with little warning, leaving resident populations poorly prepared. For a volcano to reawaken after a long phase of inactivity, the crust must rupture to provide a pathway for magma to reach the volcano. Local earthquakes can break the seal. Eruptions can pose different kinds of threat at the same time – landslides, ash clouds and fallout, fast-moving pyroclastic flows of hot gases and ash, tsunami, lava flows, earthquakes – making it especially difficult to forecast the timing, nature and impact of events.

On 13 November 1985 in the Andes Mountains of Colombia, a large stratovolcano called Nevado del Ruiz began spewing lava and ash high into the night sky. The summit crater was capped with glaciers and thick snow. Intense heat from the ash fallout rapidly melted snow and ice and the meltwater mobilised thick deposits of fine volcanic ash on the slopes of the volcano. Volcanic mudflows are called *lahars* – one of the deadliest volcanic hazards. They can reach great speeds on the steep flanks of a stratovolcano. The torrent of mud from Nevado del Ruiz overwhelmed the town of Armero, 50 kilometres from the volcano's summit, killing some 23,000 inhabitants, about 70 per cent of the population. These were the deadliest lahars in recorded history. Amero is now a ghost town of smashed and buried houses.

Volcanic activity does not have to involve mountains of ash and lava. In Iceland, for example, huge volumes of lava ooze from cracks in the ground. These fissure volcanoes can erupt for years and plaster landscapes with basalt lava that hardens to form formidably rugged terrain. By studying the lava, pumice and ash deposits from an ancient eruption, it is possible to work out its scale and intensity, and even which way the wind was blowing at the time. Techniques such as radiocarbon dating can also reveal the ages of eruptions from the distant past, enabling estimates of how often a volcano erupts. This information is vital for understanding the hazard they pose. Explosive eruptions that blast huge quantities of material high into the atmosphere are called *Plinian* eruptions after the Roman scholar Pliny the Younger, who witnessed and documented the AD 79 eruption of Mount Vesuvius that buried the Roman cities of Pompeii and Herculaneum. Both cities were engulfed by deadly pyroclastic flows.

Famously, 1816 is known as the year without a summer. The largest eruption of the past two hundred years took place in Indonesia in 1815 on the island of Sumbawa when Mount Tambora erupted catastrophically, disgorging 180 cubic kilometres of pumice and ash.

This colossal eruption ravaged the local landscape, destroying crops and pasture. Pyroclastic flows killed thousands on the flanks of the volcano and many more died in the following months as a result of starvation. The eventual death toll may have exceeded 100,000. The Tambora eruption also triggered a global climate crisis.

Sulphurous dust from the eruption circled the globe for several years, reflecting some sunlight back into space and cooling the Earth. Weather patterns were disrupted, crops failed and famine followed. There was great suffering in Europe, especially in Germany, Austria, Switzerland and France, with record numbers starving in Paris in 1816. Mary Shelley wrote *Frankenstein* on the shore of Lake Geneva in July when the 'wet, ungenial summer, and incessant rain' kept her party holed up for days. Shelley and her friends saw thick snow, thunderstorms and darkened skies. The link between her horror novel and the grim climate of 1816 has attained mythical status.

Episodes of sustained and widespread volcanism much earlier in Earth history had even greater impacts on the climate system and triggered mass extinctions. The breakup of supercontinents by rifting is accompanied by immense lava floods. The formation of extensive basalt landscapes can trigger climate cooling because chemical weathering of the basalt by rainwater sucks carbon dioxide out of the atmosphere. The weathered minerals are washed to the ocean and stored in carbonate-rich sediments.

Volcanologists have devised a scale called the Volcanic Explosivity Index (VEI) to describe the magnitude of past eruptions. This is based on eruption characteristics such as the volume of material ejected and the height of the eruption cloud. A value of VEI 0 is given to eruptions that are non-explosive. The biggest eruptions identified in the geological record are called *supereruptions* with VEI of 8. The volcanic deposits of Yellowstone National Park in Wyoming, USA, preserve evidence of three 'supereruptions' from the Pleistocene ice age, around 2.1 million, 1.3 million and 640,000 years ago. Volcanologists have calculated that the supereruption of 2.1 million years ago (VEI 8) ejected 2,500 times more ash than the Mount St Helens eruption of 1980. We do not know when the next eruption of such magnitude will strike.

# CHAPTER 20

# The Last Supercontinent and the Great Dying

At the start of the Permian, some 298.9 million years ago, it was (theoretically) possible to walk from what is now the eastern extremes of Russia to Europe, across North and South America, via southern Africa and parts of Antarctica, all the way to Australia. It would be an extremely challenging trek tens of thousands of kilometres long, traversing vast deserts, mountain ranges and tropical forests. But it was possible because all the major landmasses were joined together in a C-shaped supercontinent called Pangaea.

Supercontinents are vast ancient landmasses assembled from most or all of the existing continents. Pangaea was the first supercontinent to be recognised by geologists – though we now know it wasn't the first – and was surrounded by a vast ocean known as Panthalassa, from the Greek meaning 'all sea'. It existed from about 320 to 200 million years ago, when it began to break up at the start of the Jurassic Period. That breakup is still in progress.

It is now recognised that the assembly and breakup of supercontinents driven by plate tectonics is a fundamental process that punctuates much of Earth history. There were supercontinents before Pangaea and there will be others in the distant future. Understanding the wanderings of this gigantic geological jigsaw puzzle is important because the pattern of land and ocean across the globe exerts an important influence on all aspects of physical geography, including the direction of ocean currents and the global climate system. The location of the continents also influences where mountain ranges, ice sheets, volcanoes and earthquakes form, as well the size of river basins and deltas, and the distribution of plants and animals.

Earth history is punctuated by supercontinent cycles and there is broad agreement that three cycles of supercontinent assembly and breakup have taken place over the past 2 billion years. The history of Pangaea is the best-known because it is the most recent and its breakup is still happening. The geography of Pangaea is of particular interest because it was the only supercontinent with complex terrestrial ecosystems with forests, insects and large animals that extended from high northern latitudes down to high southern latitudes.

Climate had already become cooler and drier towards the end of the Carboniferous because the greenhouse effect weakened with the excess of oxygen in the atmosphere. Large ice sheets were still present in the southern polar lands. This drying climate was a major change from the tropical humidity that saw coal-forming swamplands dominate much of the Carboniferous Period. During the course of the Permian (the last period of the Palaeozoic Era) the climate continued to dry and vast conifer forests of seed-producing trees emerged that were better adapted to seasonal drought. As the climate warmed, the vast tropical interior became hot and dry and some animals evolved new ways of coping with the loss of wetlands.

Early reptiles had evolved from amphibians in the Carboniferous and quickly diversified, developing various adaptations to dryland habitats including moisture-retaining skin. Reptiles belong to the amniotes – vertebrates producing embryos that develop within a protective membrane. These animals could occupy a range of habitats because they did not need to lay their eggs in water. The hard-shelled egg was a key evolutionary development that allowed amniotes to dominate terrestrial ecosystems in the Permian. Other reproductive strategies were important too. Recent work has revealed that early amniotes had also evolved the ability to retain embryos inside the mother so birth could be timed to coincide with favourable environmental conditions.

The amniotes soon diverged into two major animal groups: the *sauropsids* (meaning 'lizard faces', leading to today's reptiles and birds) and the *synapsids* (named after a particular bone structure in the skull). The synapsids became the dominant animals in the Permian landscape and are important for understanding how mammals, including humans, evolved. The fossil record shows a gradual transition from reptile-like to mammal-like anatomy through changes in jaw structure, limb posture and ear bones. Synapsids have a single opening behind the eye socket on each side of the skull allowing for larger jaw muscles – improving bite strength and chewing – a trait that carried over and evolved in mammals.

*Dimetrodon* was a synapsid and is one of the most well-known animals from the Permian. Its name means 'two sizes of teeth' – it had rows of large and small teeth designed for gripping, piercing and tearing. It was a fearsome apex predator that fed on fish and small vertebrates in warm, swampy, seasonally flooded deltas. It was a big beast – the largest fossils exceed 4.5 metres in length. It is best known for the distinctive vertical arched sail protruding from its back. There has been much speculation about the purpose of this sail and whether it helped to regulate body temperature, served as camouflage or was used in courtship displays. It may have been all three.

More than a dozen species of *Dimetrodon* have been recovered from the early Permian red sandstones and mudstones of Texas and Oklahoma. The fossil skeletons of these remarkable beasts are popular attractions in many North American museums. Despite often featuring on the pages of dinosaur books, *Dimetrodon* was *not* a dinosaur. It died out about 40 million years before dinosaurs emerged. *Dimetrodon* is, in fact, more closely related to living mammals (including humans) than to any extinct or living reptile.

Its anatomy illustrates the origin of many features found in mammals, including a large brain, and a single lower jaw bone with teeth designed for chewing.

The end of the Permian Period 251.9 million years ago is marked by the most devastating extinction in Earth history that saw unprecedented losses of animals in the oceans and on land. This is the Permo–Triassic (PT) mass extinction that marks the boundary between these geological periods. The fossil record at the end of the Permian shows thriving marine ecosystems – oceans full of life – capped by an abrupt and catastrophic shift to 96 per cent of species lost. On the new supercontinent of Pangaea about 70 per cent of animal species were wiped out. This was the Great Dying. Complex life was almost annihilated. In terms of biodiversity loss it was the biggest crisis in Earth history – the very habitability of the planet may have come under threat.

The end-Permian extinction is the biggest of the five mass extinctions of the past 500 million years, and it was geologically sudden, with some estimates pointing to it being concentrated in a period of less than 60,000 years across the Permo–Triassic boundary. There has been much debate about the cause of this catastrophe. Some researchers have argued for an asteroid impact like the event that killed off the dinosaurs, but the evidence is not convincing. It is now widely accepted that volcanic activity was the trigger for a cascade of environmental changes that made life impossible for most creatures on Earth at the end of the Permian.

The volcanic activity took place on such a huge scale that it caused intense greenhouse warming that devastated planetary life support systems. In the middle decades of the twentieth century, Russian geologists in Siberia mapped basalt rocks covering an area about two-thirds the size of the United States. These, known as the Siberian Traps, are called flood basalts because they literally flooded the landscape with lava. Some of the lava flows extend for hundreds of kilometres. As one layer hardened into basalt rock, another eruption buried it with yet more lava, so the layers built up over time. In places these ancient lavas are several kilometres thick. The Siberian eruptions ejected about 3 million cubic kilometres of lava. The huge quantities of gases released by this volcanic activity changed the climate, caused acid rain and poisoned many ecosystems.

A great deal of work has now been conducted on dating the ancient lava flows in Siberia. This has shown that the volcanism corresponds with the Permian mass extinction. The colossal scale of these ancient lava flows points to a phase of volcanic activity the like of which has not been seen since. Volcanism in this region began in the late Permian and continued into the Triassic. The dates show that a major pulse of flood basalt volcanism took place at the Permo–Triassic boundary with another phase about 10 million years later.

Geologists call these regions of turbocharged volcanic activity 'large igneous provinces' (LIPs) and they can be active for tens of millions of years. Such vast and sustained outpourings of lava are fed by a phenomenon known as a mantle super plume. We do not fully understand how mantle plumes develop, but they are unusually hot upwellings of molten rock fuelled by heat from deep in the mantle – perhaps even from the core–mantle boundary. LIPs are volcanic activity in overdrive. They are a key feature of Earth history and over two hundred LIP events have been identified in the geological record. Most took place before the emergence of complex life, but it is fascinating to note that major extinction events of the last 260 million years show an almost perfect match-up with LIP eruptions.

The Siberian Traps volcanism warmed the global climate by pumping huge amounts of carbon dioxide into the atmosphere for hundreds of thousands of years. Mean global temperatures increased by at least 5 to 8 degrees Celsius. The volcanism and climate warming triggered a cascade of environmental changes that devasted ecosystems on land and in the oceans. Ecosystems such as peat-forming wetlands were eradicated. Acid rain could have been a key driver of ecosystem destruction because eruptions also pump sulphur dioxide gas into the atmosphere, creating drop-lets of sulphuric acid. High concentrations of sulphur dioxide also create a volcanic smog, making it difficult to breathe.

Volcanic eruptions produce gases called halocarbons that destroy the ozone layer, thereby exposing Earth's surface to lethal levels of ultraviolet radiation. Trees can become sterile when exposed to high levels of UV radiation because seed cones and pollen grains are damaged. While there are studies reporting charred plant remains in fossil forests that point to an increase in wildfires at the time of the extinction event, there is some uncertainty about how extensive the losses were among plant species. Plants can have greater resilience to environmental stresses because seeds and root systems can remain viable for decades.

Rapid global warming set in motion a cascade of impacts such as ocean heating and acidification. When the oceans become more acidic it becomes more difficult for animals to make carbonate shells and corals bleach and die. Ocean warming in the polar regions may have released stores of methane gas from beneath the sea floor, which intensified the greenhouse conditions.

When the oceans get warmer the amount of free oxygen seawater can hold decreases, so there is less oxygen available for marine life. All animals need oxygen to breathe, and zooplankton at the bottom of the food chain can be especially sensitive to changes in oxygen concentration. They can move to cooler and deeper waters, but food may be scarce there and reproduction more difficult. Many higher-order species of fishes, squids and whales eat zooplankton or eat creatures that feed on zooplankton. If the bottom of the food chain is wiped out, the consequences for the rest of the food web are catastrophic. If warming kills off the tiny marine creatures that perform photosynthesis, there is less food and less oxygen.

A rapidly warming ocean is a potentially deadly scenario for marine life because even minor falls in oxygen can affect the viability of species. There is a double whammy here because as water temperatures go up animals use *more* oxygen, so an extreme ocean heatwave can lead to mass deaths. Some creatures can migrate to cooler waters by heading poleward or finding refuge in deeper parts of the ocean. But when warming is extreme, global and prolonged, these options are cut off and the impacts can be devastating. During the mass extinction at the end of the Permian vast tracts of the ocean became oxygen-starved dead zones.

The Great Dying marked the end of the Palaeozoic, an era that had begun with a burst of new lifeforms in the Cambrian Period.

At the close of the Permian, the largest synapsids, including *Dimetrodon*, became extinct, along with many sauropsids and amphibians, with herbivorous groups being hit the hardiest. Insects endured the greatest species losses in their history, with the loss of several entire orders. Among the casualties in the oceans were two kinds of reef-building coral that had first appeared in the Ordovician, sea scorpions and all the remaining species of trilobites. Deep-water sponges saw large species losses, as did brachiopods, crinoids, gastropods and ammonoids. Some species of shark survived the extinction event in deep ocean refuges.

The mass extinction at the end of the Permian shows that the habitability of our planet is not guaranteed. We still have much to learn about the impacts of a runaway greenhouse climate and how quickly life support systems could collapse. But life did recover. There were refuges in the oceans and on the Pangaea supercontinent where species were able to survive the traumas triggered by the volcanic eruptions.

The northern coast of Pangaea was dominated by a river delta more than 1,000 kilometres across that covered an area of about ten times the size of the modern delta of the Amazon. This Permo–Triassic delta was built up by sediments and floodwaters transported during heavy monsoon rains. Because delta wetlands provide a rich mix of ecosystems, this vast northern delta complex may have provided the best conditions for the resurgence of life in the Triassic Period. The physical geography of the supercontinent produced a super-sized delta where life could flourish after the catastrophe of the end-Permian extinction.

The biosphere teetered on the brink some 252 million years ago at the close of the Palaeozoic Era when complex life came closest to being wiped out, but about 30 per cent of land animal species survived the mass extinction. These animals, including the sauropsids, were able to exploit new ecological opportunities and diversify in the Triassic. The sauropsids evolved to include the entire dinosaur lineage (which includes all birds), as well as lizards, snakes, turtles and crocodiles. It was the dinosaurs that dominated the next great era of Earth history.

# CHAPTER 21

# The Age of the Dinosaurs

About 20 million years after the greatest mass extinction in Earth history, the first dinosaurs emerged in the middle of the Triassic Period (251.9 to 201.4 million years ago). They evolved into a spectacularly diverse group of animals in the Jurassic and Cretaceous periods and came to dominate land-based ecosystems for more than 140 million years. These three geological periods form the Mesozoic Era, often called the age of reptiles.

The scientific study of dinosaurs can be traced to the early decades of the nineteenth century. A dinosaur was formally named for the first time in 1824 by the Oxford University geologist and priest William Buckland (1784–1856), who studied the massive bones and teeth of *Megalosaurus* that had been found in local limestone quarries. *Megalosaurus* means 'great lizard' and its discovery caused a sensation. In 1842 the naturalist Richard Owen (1804–1892) proposed the name Dinosauria, meaning 'terrible lizards', to describe this new order of reptiles. Dinosaurs have been a source of wonder for every generation since.

When I was about six or seven my big brother received a book about dinosaurs for his birthday. We learned the names: *Stegosaurus, Allosaurus, Brontosaurus, Triceratops, Barosaurus, Diplodocus, Tyrannosaurus rex* and so on. We pored over the illustrations of these weird and wonderful giants – carnivores and herbivores, predator and prey. Their habitats were lush tropical forests and hot desert landscapes. The dinosaurs were coloured grey, brown or green and had dull, scaly skin. We learned they were sluggish, cold-blooded creatures with tiny brains. The book was rather vague on why the dinosaurs died out, simply stating that they had failed to adapt to a changing environment. This was dinosaurs for kids in the 1970s.

Many extraordinary discoveries over the past few decades have radically changed our understanding of what dinosaurs looked like, how they behaved, the environments in which they lived and how they became extinct. Modern research places as much importance on retrieving details about the ancient environment and the ecology of the animals as it does on recovering the bones. Dinosaur fossils have been found on every continent, including Antarctica, but recent discoveries in China have been especially influential because burial conditions have allowed the preservation of extraordinary detail. Well over 2,000 species of extinct dinosaur have now been named, and this figure increases each year. When Richard Owen coined the term Dinosauria, there were just three.

From about the 1870s to the 1930s there was a heroic age of dinosaur fossil hunting in the United States, when exciting discoveries were frequently reported in the newspapers. Big city museums backed by wealthy industrialists fuelled the search for new dinosaurs. Fossil hunters became celebrities and fossil hunting became territorial and fiercely competitive. Most of the classic dinosaur species in my brother's book were discovered and named by just two men – Edward Drinker Cope (1840–1897) and Othniel Charles Marsh (1831–1899). They engaged in a spectacularly bitter feud and behaved appallingly. Their decades-long rivalry became known as the Bone Wars. Today dinosaur palaeontology is (mostly) much more collaborative.

It is important to appreciate that many dinosaur species were separated by geography and great spans of time. Dinosaurs emerged in the Triassic Period on the supercontinent of Pangaea, but the best-known dinosaurs come from the Jurassic (201.4 to 143.1 million years ago) and Cretaceous (143.1 to 66 million years ago) periods, when new oceans were forming as Pangaea fractured into multiple continents. *Stegosaurus*, for example, lived in the Upper Jurassic between about 155 and 143 million years ago, while *T. rex* did not appear on the scene until the later part of the Cretaceous some 80 million years later. The length of time *between* these iconic dinosaurs is greater than the time interval between *T. rex* and humans.

The world was changing in the Mesozoic, the first 10 million years of which saw the Earth recovering from the devastation of the end-Permian extinction. Global temperatures in the Lower Triassic were much hotter than today and may have been the warmest of the last half-billion years, a legacy of the huge volumes of greenhouse gases pumped into the atmosphere by the Siberian Traps volcanoes. The polar regions were ice-free with lush vegetation on the southern and northern margins of the Pangaea supercontinent. Conifer forests became widespread, but the vast interior of Pangaea was a bone-dry hot desert. The close of the Triassic around 201 million years ago saw another mass extinction which mainly affected large amphibians, marine reptiles and many species of fish, corals and molluscs. Volcanism was again the likely culprit when an intense phase of flood basalt eruptions known as the Central Atlantic Magmatic Province (CAMP) took place. Lava poured out over vast areas in the early stages of the formation of the Atlantic Ocean as the supercontinent Pangaea began to break apart. These lavas covered an area roughly the size of today's North America. Again, volcanic activity released enormous amounts of carbon dioxide and sulphur dioxide, leading to rapid global warming, ocean acidification, and oxygen depletion in deeper ocean basins. On land, dinosaurs and early mammals survived these environmental changes.

At the University of Manchester we are lucky to have a museum on campus with a natural history gallery that tells the story of life on Earth. The star of this collection is Stan the *Tyrannosaurus rex*. Stan is named after Stan Sacrison, the amateur palaeontologist who discovered the skeleton, but we don't know if this *T. rex* is male or female; many dinosaur skeletons do not show clear physical differences between the sexes. Stan is a cast of one of the most complete *T. rex* skeletons ever found. The bones record the scars of ancient battles: there is a neck fracture, a head injury at the back of the skull with a cavity the same size as a *T. rex* tooth, some broken and healed ribs, and scars on the cheeks. More casts have been made of this dinosaur skeleton than any other, so Stan may well be the most viewed dinosaur of all time. For a while Stan was also the most expensive. On 6 October 2020 Christie's auction house in New York sold the original Stan for almost $32 million.

*T. rex* belongs to the group of dinosaurs known as theropods, meaning 'beast-footed'. Most theropods were carnivores with sharp, blade-like, curved teeth for ripping flesh and prominent claws for trapping prey. There has been a huge amount of work on theropod evolution and hundreds of fossil species have been named.

In 1861, just two years after Darwin published *On the Origin of Species*, one of the most iconic dinosaur fossils was discovered in a limestone quarry in Bavaria. The therapod fossil was named *Archaeopteryx*, combining two Ancient Greek words that mean 'ancient' (*archaīos*) and 'feather' or 'wing' (*ptéryx*). The now-famous slab of limestone was purchased by London's Natural History Museum. *Archaeopteryx* is a transitional fossil that has created excitement, confusion and controversy in equal measure. It offered powerful support to Darwin's theory of evolution.

*Archaeopteryx* was a small theropod with an unusual mix of features. It was part bird, part reptile. It had a small body about the size of a magpie, with feathered wings and legs, and feet with some avian aspects but similar to the foot of a raptor dinosaur. It also had jaws with sharp teeth, a long bony tail and the hands of a reptile. Palaeontologists think that *Archaeopteryx* was able to flap its wings and fly short distances. We cannot be sure that *Archaeopteryx* was the first bird – it was a bird-like dinosaur that lived towards the end of the Jurassic about 150 million years ago. This fossil is hugely important because it provided the first piece of evidence to show that birds are direct descendants of dinosaurs. It busted the myth that all the dinosaurs became extinct – one lineage of dinosaur developed the ability to fly and survived the mass extinction at the end of the Mesozoic Era. This group evolved into the birds and diversified spectacularly (Chapter 23).

The Lower Cretaceous rocks of China have yielded beautifully detailed fossils that show the widespread occurrence of feathers in dinosaurs large and small. This revelation transformed our understanding of what many dinosaurs looked like. Some dinosaurs were completely covered in simple hair or brush-like feathers, others had clusters of feathers on their legs or back. These fossils from China show that more primitive groups of non-flying dinosaurs first evolved feathers for reasons unrelated to flight, though the fossils themselves are not from a time before flight. Many palaeontologists now think that feathers played a key role in camouflage and sexual selection. Did some dinosaurs strut around like peacocks? Did they dart around noisily like giant turkeys?

Melanosomes are tiny, pigment-filled structures found inside cells. They give colour to skin, feathers, hair and scales. Under the right conditions they can fossilise and survive for millions of years. By comparing the colours of modern birds with the melanosomes in their feathers, predictions can be made about the plumage colours of extinct dinosaurs in cases where the feathers are very well preserved. When researchers analysed fossilised melanosomes in *Archaeopteryx* feathers, they predicted with 95 per cent probability that its feathers were black.

A new dinosaur, *Yutyrannus huali*, was described in 2012. Its Latin and Mandarin name means 'beautiful, feathered tyrant'. An adult and two juveniles were excavated in Liaoning Province in northeast China. The bodies were covered in fine, downy feathers, presumably to help keep them warm. The adult would have weighed more than a tonne. This giant carnivore is the largest creature with full plumage so far discovered. This is just one of the recent findings that have fundamentally changed our understanding of the appearance of dinosaurs.

If you compare the appearance of the dinosaurs in the original *Jurassic Park* (1993) movie with *Jurassic World Dominion* (2022), some of the key discoveries made in the three decades between these films become apparent. In 1993 the raptors had scaly brown-green skin without a feather in sight. By 2022 an adult raptor sported a spectacular plumage of large feathers coloured deep red and grey. While this is in line with the latest science, we should note that an adult *Velociraptor* stood about half a metre high at the hips, about the size of a Labrador. Hollywood trebled their size in both movies to make them more terrifying!

Dinosaurs had a hip structure that gave them straight back legs and an upright stance. The study of dinosaur trackways and comparisons of fossil anatomy with modern animals has revealed valuable information about how dinosaurs moved. Modern-day therapods like ostriches can reach speeds of more than 60 kilometres per hour. We know from fossil trackways preserved in Cretaceous rocks of northern Spain that some therapod dinosaurs could reach speeds in excess of 45 kilometres per hour. These tracks also indicate impressive agility – these animals could suddenly reduce their speed before accelerating rapidly. There has been much debate about how fast *T. rex* could run. There is a famous scene in *Jurassic Park* where a *T. rex* chases a jeep at top speed. The latest research suggests that an adult *Tyrannosaurus* was, at best, a reluctant jogger with limited, if any, running ability. The fastest dinosaurs were medium-sized animals.

The sauropods were a group of long-necked, plant-eating dinosaurs that grew to an enormous size and dominated Jurassic and Cretaceous ecosystems. *Brontosaurus*, *Brachiosaurus* and *Diplodocus* are some of the best-known sauropods. Some reached lengths of 20 to 30 metres and their weight exceeded 50 tonnes. They were the largest animals to ever walk the Earth and there has been much debate about how they moved and their social behaviour. A common view in the early twentieth century was that these beasts must have been aquatic because they were too big to support their own weight on land. We now know this was not the case but there is still much to learn about the way they moved. Some researchers have argued that they formed large herds that were segregated by age. Trackway evidence has been interpreted as showing that juveniles formed groups that were separate from those of adults.

Not all dinosaurs lived in the warm tropics. The polar regions were also important dinosaur habitats – therapod trackways have been found in the Denali National Park in Alaska, for example. The beast that made these footprints was a cousin of *Therizinosaurus*, which means 'scythe lizard'. It was a strange-looking creature – a feathered, 6-tonne theropod with a giraffe-length neck, small head and chunky lower body. This dinosaur had huge, elongated talons like the scythe of the grim reaper. Even more strangely, it appears to have been a carnivore that turned vegan. Trees and shrubs were widespread in the warm Cretaceous climate, and a big beast like *Therizinosaurus* would have consumed copious amounts of plant matter every day. Its claws were probably used machete-style for cutting stalks and branches.

The ecology of polar dinosaurs has generated a great deal of interest. Trackways from groups of both adult and juvenile dinosaurs have now been found in Alaska. The presence of young animals is a strong indication that they spent all year in the far north and were not just spring and summer visitors. At the Colville river site in northern Alaska, which dates to the late Cretaceous, palaeontologists have found fossils of several species of dinosaur, both herbivores and carnivores as well as baby dinosaurs. Dinosaurs were nesting in the Arctic. The polar regions were much warmer than today during the Cretaceous but to cope with the colder, dark winter months some of the smaller dinosaurs may have hibernated, while the larger beasts may have built up layers of fat to keep warm.

The discovery of polar dinosaurs has reopened the long-running debate about whether dinosaurs were cold- or warm-blooded. Even in the Cretaceous greenhouse, dinosaurs that lived in the polar regions year-round must have been able to generate at least some of their own body heat. Feathers were probably important for heat retention.

There is compelling evidence from Upper Cretaceous rocks in the Gobi Desert that dinosaurs nested in colonies and took care of their eggs. Palaeontologists have established from the beautifully preserved nest sites that some 60 per cent of the eggs hatched successfully – a success rate that compares with modern birds and reptiles who protect their eggs. There is also evidence that some dinosaurs were smart enough to return to the same nesting sites year after year. This complex community behaviour was in place in the early Jurassic. These insights into dinosaur behaviour form part of a growing body of evidence that refutes the long-held belief that dinosaurs were brutish giants who lacked the intelligence to care for their offspring.

The way we think about dinosaurs has changed radically in the last few decades. CT scans of dinosaur skulls show that their brain cavities were quite large, so the classic trope of slow-witted dinosaurs is wrong. Many aspects of dinosaur behaviour gleaned from the fossil record can be seen in animals today. Perhaps they are not such distant relatives after all.

# CHAPTER 22

# Mary Anning's Jurassic World

When dinosaurs dominated the Earth the oceans were full of weird and wonderful sea monsters. The Mesozoic Era saw the emergence of a diverse and highly successful group of marine reptiles that included perhaps the most ferocious marine predator in Earth history.

Some of the most important specimens from the Jurassic ocean were discovered in the early decades of the nineteenth century by a young woman from a poor family who had no formal education. Against all the odds, she gained a reputation as the finest 'fossilist' of her generation, known to all the leading collectors in Britain. Not only was her expertise sought by the foremost geologists of the day, her discoveries also provided hard scientific evidence about ancient life, extinction and deep time, challenging conventional views about the history of the Earth.

Mary Anning (1799–1847) was born in Lyme Regis on the Dorset coast of southwest England on 21 May 1799. Out of nine children only Mary and her brother Joseph (1796–1849) survived to adulthood. She narrowly escaped death herself at the age of fifteen months when taken to shelter from a thunderstorm beneath a tree. The three women with her were killed stone dead by a lightning bolt.

Lyme Regis became a fashionable seaside resort in the early nineteenth century when the Napoleonic Wars deterred travel to the Continent. Many visitors were keen to add fossils to their cabinets of curiosities. Mary's father Richard was a cabinet maker who showed his children how to retrieve fossils from the local cliffs and beaches. In 1810, when Mary was just eleven, he died, leaving the family destitute and dependent on local parish relief. To make ends meet, Mary searched for fossils with her brother and sold them to tourists from a table outside their house. Winter was often the best time to find fossils after storm waves had crashed against the soft cliffs and created fresh rockfalls. Mary Anning spent long hours scouring the coastline in all weathers. It was a hard and precarious business.

The places where she searched for fossils are now part of the Jurassic Coast UNESCO World Heritage site, stretching for 95 miles along the coasts of Dorset and east Devon in southwest England. The rocks along this stretch of coast span all of the Mesozoic Era (Triassic, Jurassic and Cretaceous periods), recording 185 million years of Earth history. This majestic coastline also includes textbook examples of classic landforms, including rock arches, stacks and landslides. The Jurassic Coast attracts visitors from across the world and is a popular location for Earth science and physical geography field trips. Mary Anning's story and scientific legacy are major contributors to its popularity.

Many of the fossils Mary Anning found came from the Blue Lias, a Jurassic rock formation exposed on the Dorset coast. It is a sequence of alternating hard limestones and softer mudstones. Each bed was once a layer of soft mud at the bottom of a deep tropical ocean, when the slab of crust that is now Britain lay just north of the equator. The fine muds provided ideal conditions for preserving fossils of the animals that lived in this sea, especially if they were buried rapidly.

In 1811 Joseph Anning found a huge skull with eye sockets the size of dinner plates and long crocodile-like jaws packed with conical pointed teeth. The following year Mary located the body of this beast and carefully excavated its bones. It was the most complete skeleton of an ichthyosaur anyone had ever seen. When skull and body were reunited, the ancient creature was some 17 feet long. Although ichthyosaur means 'fish lizard', it was neither fish nor lizard; neither was it a dinosaur. Some of the early collectors called ichthyosaurs sea dragons to generate publicity and no doubt raise their prices. Ichthyosaurs were marine reptiles and important predators in Mesozoic marine ecosystems. Another well-preserved specimen discovered by Mary Anning in the Jurassic cliffs even has fish bones from its last supper lodged inside its ribcage.

In December 1823, Mary discovered the first complete skeleton of a plesiosaur with almost all the bones present and in their correct anatomical position. Some bones of this marine reptile had already been found, but this was the first time the whole skeleton could be studied. With its neck as long as its body, no one had ever seen an animal quite like it.

Plesiosaurs were carnivorous marine reptiles, first appearing in the Triassic. They spent their entire life cycle in the water and gave birth to live young. Like ichthyosaurs, they were air-breathing so must have surfaced frequently. With their small heads, long necks (some species had necks that were 7 metres long) and barrel-shaped bodies with four flippers, there is nothing quite like them swimming in today's oceans. Little wonder this creature may have been the inspiration for the Loch Ness Monster. The colourful Oxford don William Buckland likened it to 'a sea serpent run through a turtle'.

Mary Anning's 1823 plesiosaur with its astonishingly long neck was hugely controversial because some of the leading palaeontologists of the day could not believe that it was real. Even the brilliant French anatomist Georges Cuvier (1769–1832) thought it must have been a mix of bones from different animals. The specimen was examined closely at a meeting of the Geological Society and drawings were sent to Cuvier, who was soon convinced that the specimen was genuine.

Even though she was excluded from them, Mary Anning's reputation grew quickly in the rarefied circles of London's scientific
societies. She often corresponded with aristocrats and her best
finds commanded good prices. In 1830 she discovered a new
species of plesiosaur – a beautifully preserved juvenile specimen,
ideal for a Georgian drawing room. She sold it for the huge sum of
over £200 to Lord Cole, who had one of the finest collections of
fossil fish in Europe.

But Mary Anning was much more than a seller of fossils. And
she had no time for the fancies practised by other fossil sellers such
as carving snake heads on their tightly curled ammonites to
enhance their mystique and increase their value. Fossils had long
been a rich source of superstition and folklore, but for Anning they
were keys to unlock scientific understanding. She understood the
anatomy of these ancient animals and was a highly skilled exca-
vator and fossil preparator. She became an expert in the identification of ammonites and even the fossil faeces from marine reptiles.
And she also developed a keen understanding of the geological
context of her finds and how the animals became fossils.

Mary Anning shared her wisdom in the field with the finest
minds in geology. By the time she was in her twenties, Mary had an
international reputation as a 'fossilist'. One aristocratic visitor to
Lyme Regis noted in her diary that twenty-five-year-old Anning
was 'in the habit of writing and talking with professors and other
clever men on the subject, and they all acknowledge that she
understands more of the science than anyone else in this kingdom'.
The Swiss naturalist Louis Agassiz (1807–1873), who famously
pioneered the idea of a great ice age (Chapter 31), visited Mary in
Lyme Regis and named two species of fossil fish after her.

In 1828 Mary Anning excavated a pterosaur, the first specimen
of this flying reptile to be found in Britain. Pterosaurs are found in
the fossil record from the end of the Triassic to the late Cretaceous.
The largest had a huge wingspan approaching 10 metres and could
fly very long distances: they were the first creatures capable of
intercontinental flight. The species Mary Anning found was
*Dimorphodon*, which lived about 200 million years ago in the very

early Jurassic. Pterosaurs are not ancestors of today's birds, nor are they flying dinosaurs. They are a distinct group of flying reptiles that became extinct at the end of the Cretaceous.

Henry De la Beche (1796–1855) was an important figure in nineteenth-century geology. His family moved to Lyme Regis when he was a teenager and he became friends with Mary Anning, often accompanying her on fossil-hunting walks. De la Beche had inherited wealth that allowed him to pursue geological interests. He was educated, middle class and well connected. As a Fellow of the Geological Society, he reported finds from the Jurassic Coast at its meetings in London. He published several books on geology and was a gifted illustrator. His most famous artwork, *Duria antiquior – a more ancient Dorset*, was inspired by Mary Anning's discoveries. It is the first example of a genre that has become known as 'palaeoart'. The action-packed cartoon shows a Jurassic ocean stuffed with marine reptiles. The scene is dominated by a fearsome ichthyosaur with the lanky neck of a plesiosaur clamped in its jaws, but there are also ammonites and fish and even fossils forming on the seabed. Above the water, pterosaurs flap around in the skies while a crocodile and turtle survey the scene from the shore beneath tropical vegetation. It is a quite brilliant reconstruction of a long-vanished world. De la Beche sold prints of *Duria antiquior* to help Mary pay her bills.

The Jurassic Coast continues to be a treasure trove for exceptional fossils. In early April 2022, the snout of a giant pliosaur with interlocking teeth was found on the beach near Kimmeridge in Dorset. It had fallen 15 metres down the cliff. After a drone survey of the cliff face located the rest of the massive skull, it was painstakingly excavated by a team suspended precariously from the clifftop on climbing ropes. What would Mary Anning have made of all this? This specimen has generated huge excitement because it is the most complete and well-preserved pliosaur skull ever discovered. The skull is 2 metres long and every bone is present.

The pliosaur, which is a type of plesiosaur, was a ferocious apex predator, the largest marine reptile ever to live and the biggest creature in the Jurassic ocean. The powerful jaws of the Kimmeridge specimen contained 130 crocodile-like teeth and has been nick-named Sea Rex! Saltwater crocodiles have the most powerful bite force of any living animal – their jaws snap closed with astonishing power and violence. The bite of a huge Jurassic pliosaur would be twice as powerful as the saltwater crocodile. It was the most fero-cious killing machine in the Jurassic ocean.

When Mary Anning was born, most people in England got their Earth history from the Bible. During the course of her lifetime, geology and palaeontology became established as modern sciences. Her discoveries in the Jurassic rocks of Dorset were hugely influ-ential as geologists were grappling with radical new ideas about deep time that challenged established wisdom. The extinct animals that she presented to the world played an important part in making geology a rational objective science.

The remarkable fossils Mary Anning discovered are still studied by palaeontologists today. Some of her most famous finds are on display at the Natural History Museum in London and are seen by millions of visitors every year. It is difficult to overstate her legacy. Mary was highly regarded for her skills and knowledge, but her social status, gender, perilous finances and remote rural location meant that she could never play a full role in the London-based science debates and so her practical and intellectual contributions were not fully recognised. Many others were similarly excluded at that time, but it certainly did not mean she was not valued and respected in the geological community. When Mary was defrauded of her life savings in 1833, her friends at the Geological Society successfully lobbied the prime minister, Lord Melbourne, to grant her a small pension.

Many Anning died in 1847 after a painful battle with breast cancer. In 1850 her life was commemorated by the installation of a stained-glass window in St Michael's church, funded by members of the Geological Society and the local vicar. This pretty limestone church sits on the cliff top in Lyme where Mary is buried with her brother Joseph. If you stand by their gravestone, you can hear the waves breaking on the foreshore below.

The significance of Mary Anning's work is now widely recognised, and she is rightly lauded as a trailblazer of palaeontological theory and practice. She is championed as a pioneering woman of science whose extraordinary gifts and work ethic made a difference in an era when the odds were heavily stacked against her. In 2022 a superb bronze statue of Mary was unveiled in Lyme Regis, the culmination of a campaign begun by Dorchester schoolgirl Evie Swire to create a permanent and prominent memorial in Anning's home town. The statue shows Mary striding decisively towards the Jurassic cliffs, geological hammer in hand, collecting basket over her arm, accompanied by her faithful terrier Tray. The hem of her skirt is decorated with ammonites.

# CHAPTER 23

# The End of the Dinosaurs

In the summer of 1980 four scientists from the University of California published a sensational paper that transformed our understanding of the history of life on Earth. They claimed that catastrophic environmental changes triggered by a giant asteroid smashing into Earth caused the mass extinction that wiped out the dinosaurs. Their explanation for the asteroid impact some 66 million years ago was grounded in hard scientific evidence, but it was hugely controversial and dismissed out of hand by many Earth scientists. The extinction of the dinosaurs had been a mystery for the best part of 150 years, but for much of this time geology had pushed back on catastrophic explanations for events in the geological record. Had the mystery finally been solved?

The key thinkers behind this bold new theory were a geologist, Walter Alvarez (b. 1940), and his father, Luis Alvarez (1911–1998), a Nobel Prize-winning physicist who had worked on the Manhattan Project during the Second World War. On 6 August 1945 Luis Alvarez flew alongside the *Enola Gay* and witnessed the devastation of Hiroshima. His task was to measure the strength of the blast from the atomic bomb. Alvarez senior knew all about catastrophic explosions.

In the 1970s Walter Alvarez had spent several field seasons in the Apennine Mountains of central Italy studying the limestone rocks in the Bottaccione Gorge near the sleepy town of Gubbio. These rocks were laid down in an ancient tropical ocean called Tethys, the forerunner of today's Mediterranean. The sediments laid down on this Cretaceous sea floor have been uplifted to more than 500 metres above sea level by the crust-buckling northward creep of the African plate.

Alvarez became intrigued by a thin layer of dark clay within a much thicker sequence of otherwise pale limestones. The limestones below the clay were packed with shells of assorted sizes, including tiny fossil plankton. This painted a picture of a diverse and thriving marine ecosystem. The limestones immediately above the clay were almost entirely devoid of life. This stack of rocks in the middle of Italy records an unusually sudden ecosystem collapse on one small part of an ancient ocean floor, which Alvarez knew was part of the death spiral at the end of the Cretaceous Period. He became convinced that the thin band of dark clay was key to unravelling one of the greatest puzzles in Earth history. It was a singular feature just a centimetre thick and there was nothing else like it in all the hundreds of metres of Italian limestones he had observed. Where did the clay come from? What secrets did it hold about the mass extinction that not only wiped out the dinosaurs but also killed off 75 per cent of life on Earth?

The extinction at the end of the Cretaceous is one of the big five extinctions in Earth history. The Cretaceous was followed by the Palaeogene Period so this is often called the K–Pg extinction (the K comes from the German *Kreide*, meaning 'chalk'; *Kreidezeit* is the German name for the Cretaceous). The K–Pg extinction saw the death of all apex predators on land and in the oceans. All the flying reptiles perished too. In the oceans, more than 90 per cent of plankton species were wiped out, leading to a complete breakdown in food webs and ultimately mass deaths of ammonites and many fish species. Some species of shark survived, but the giant marine reptiles Mary Anning had excavated disappeared from the fossil record. The age of large reptiles on land and in the oceans ended very abruptly.

By the end of the Cretaceous Period the global geography of land and ocean was getting close to what we see today, although there were still major differences. While the Atlantic Ocean divided Europe from North America, Australia and South America were still connected to Antarctica in a great southern landmass. North America was split in half by a shallow sea and would not be joined with South America for another 60 million years. Dinosaurs were present on every continent.

The Cretaceous greenhouse climate was one of the warmest periods in Earth's history. What is now Greenland and the Antarctic were landscapes of lush temperate woodland and wetlands. Because the Earth was largely ice-free, sea level was more than 150 metres higher than today. The warm oceans were highly productive and teemed with life. The iconic White Cliffs of Dover on the southeast coast of Britain are formed of chalk laid down during the late Cretaceous from billions of tiny plankton shells.

Walter Alvarez discussed the Gubbio sequence with his father and they began to investigate the composition of the dark clay. When they looked at the very heavy platinum group elements they detected a huge spike in the rare metal iridium. This metal is denser than gold and extremely rare in the Earth's crust. It is found in much higher concentrations in meteorites. Why did this thin layer of clay contain tiny fragments of space rock? Father and son became convinced this clay was the debris from an asteroid impact – the fallout from a catastrophic explosion at the end of the Cretaceous. They found the same iridium spike at a site in Denmark precisely at the Cretaceous–Palaeogene boundary. The only explanation for this iridium anomaly, they concluded, was an asteroid strike of such magnitude that it blasted vast amounts of debris around the world.

Luis and Walter Alvarez constructed an elegant theory to explain the iridium spike and the global distribution of the clay layer. They also came up with an explanation for the mass killing of dinosaurs and marine reptiles. They calculated the asteroid was about 10 kilometres across and would have generated close to a billion times the energy of the atomic explosions that devastated Hiroshima and Nagasaki. The impact would have obliterated the asteroid and blasted hundreds of billions of tonnes of fine dust and sulphur particles from the crater site into the atmosphere. A thick cloud of dust and soot would have circled the globe, blocking out the light and heat from the Sun and initiating a deadly period of global cooling. Models of the impact and its aftermath suggest this state of freezing darkness could have persisted for several years. It would have shut down photosynthesis with deadly consequences, killing off plants on the land and plankton in the oceans. According to the Alvarezes, this crisis saw the complete breakdown of food webs and ecosystems – a global kill mechanism for large vertebrates on land and in the seas.

We can't be certain about which was the largest dinosaur, but all the candidates belong to a subgroup of sauropods called the titanosaurs. The titanosaurs were the last of the sauropods and they lived in the Cretaceous. In 2023 I saw the titanosaur *Patagotitan mayorum* on display at the Natural History Museum in London. It was one of the largest animals to walk the Earth, four times heavier than an adult *Diplodocus*. I had my photograph taken next to its thigh bone, which was 2.38 metres long! The largest dinosaurs had huge appetites. These 50-tonne beasts would have consumed hundreds of kilograms of plant matter every day. If the giant sauropods were starved of plants, the carnivores would be starved of meat.

Since the pioneering work of Alvarez and Alvarez, hundreds of sites across the world have been shown to contain the iridium spike where rocks lie at the K–Pg boundary. This bolstered the asteroid impact theory and the idea that debris and soot from the impact shrouded the entire Earth. A key breakthrough came in 1991 when a massive impact crater that dated to the end of the Cretaceous was discovered off the Yucatán Peninsula in Mexico. The Chicxulub crater is over 180 kilometres across but had been obscured from view by millions of years of sediment deposition on the sea floor.

Traces of this crater were first uncovered in the 1960s during geological prospecting by a Mexican oil company but its full significance was only realised a decade after the impact theory was published. It is one of the largest impact craters known from the last billion years of Earth history.

The impact itself would have wiped out practically all life within a 1,000-kilometre radius of the Yucatán Peninsula. The impact generated huge shock waves, extreme heat, wildfires and tsunamis that would have devastated life in the region. Species loss would have taken rather longer with increasing distance from the impact site as animals and plants were hit by the plunge in temperatures that followed.

The Chicxulub crater contains lumps of rock known as *tektites*, from the Ancient Greek *tēktós*, meaning 'molten'. When a major asteroid strikes a planetary surface, superheated material is ejected high into the atmosphere where it cools to form glassy tektites ranging in size from sand grains to small pebbles. The smallest glassy globules can be blasted thousands of kilometres from the impact site. These provide important clues about impact events, even when the craters have been buried or eroded away. Geologists have dated rocks from the base of the crater that melted during the impact. These rocks have yielded ages that match the age of the iridium spike. Tiny glassy spherules from the impact site at Chicxulub have been found in the clay layer at Gubbio in Italy.

During the formation of an impact crater, the rocks and minerals blasted from the site are subjected to enormous stresses. They are pulverised, melted, fractured and vaporised. In the case of quartz, its characteristic crystalline structure can be deformed under intense pressure to create shocked quartz. These quartz grains are common in impact craters and in the dust that settles after nuclear explosions. The discovery of shocked quartz grains in the iridium-rich clay layer at the K–Pg boundary strengthened the case for a giant impact.

The K–Pg boundary is a uniquely well-defined boundary in the geological record because the iridium spike provides a worldwide time marker. When geologists locate the clays with the iridium spike they know this layer is the same age wherever it is found. The most recent dating places this layer and the impact at 66.043 ± 0.011 million years ago.

Volcanic activity may also have played a role in the demise of the dinosaurs. Some Earth scientists even argue that the climatic impact of volcanic activity was more important than the asteroid impact. We know that large-scale volcanic eruptions were deadly in previous mass extinctions, and it is important to recognise that a phase of massive volcanic activity in India that produced the Deccan lava flows coincided with the end of the Cretaceous. Multiple volcanic outbursts over some 1.5 million square kilometres of today's west-central India may have suffocated lifeforms by contaminating the atmosphere with sulphur dioxide.

The asteroid impact (and its aftermath) is the favoured kill mechanism for most scientists who work on this problem, but some also argue for a role for the Indian volcanoes. Indeed it may have been a combination of these events. Since the Chicxulub impact is so precisely correlated in time to the mass extinction, it is hard to argue it is not the main trigger. But even Walter Alvarez has conceded that the vast scale of the volcanic eruptions in India may have played a role in the extinction. It is likely that multiple factors combined to kill off the dinosaurs, but most agree the primary cause was the asteroid impact and its aftermath. The volcanic activity had been in progress for millions of years before the asteroid struck.

There is a lively debate about the diversity of dinosaur species in the lead-up to the asteroid impact. Was diversity still high with many thriving populations or were these giants already in steady decline? This is still an open question but there is little doubt that hundreds of species of dinosaur, including the iconic species *Triceratops, Tyrannosaurus rex* and the armoured dinosaur *Ankylosaurus*, were casualties of the K–Pg extinction.

The asteroid impact did not kill off all the dinosaurs. It exterminated all the *non-avian* dinosaurs. A few species of small, feathered dinosaurs survived into the Cenozoic Era. Even the birds came close to annihilation: the latest genetic and anatomical research suggests only a handful of species survived the extinction and these were primarily ground-dwelling birds. Forests in the region of the impact site were decimated by acid rain and wildfires and by the prolonged darkness of the global dust cloud, meaning tree-dwelling birds had nowhere to roost. The clutch of lucky survivors went on to diversify spectacularly, with more than 11,000 species of birds living today. But mammals were the big winners. These small animals had kept a low profile for millions of years, but once the dust had settled on the K–Pg extinction, the age of mammals began and life flourished anew.

In the last few decades, the town of Gubbio has become a site of pilgrimage for Earth scientists and dinosaur enthusiasts where, at the famous section Walter Alvarez recorded, you can touch the dust that killed the dinosaurs.

# CHAPTER 24

# The Idea of Extinction

Eldey island is an anvil-shaped lump of volcanic rock about 7 nautical miles off the southwest coast of Iceland. It is the site of a mid-nineteenth-century extinction. On 3 June 1844, during the struggle to throttle its parents, the last viable great auk egg was crushed beneath a sailor's boot.

The great auk (*Pinguinus impennis*) was a large flightless seabird with distinctive black-and-white plumage that inhabited the wild waters and remote rocky islands of the North Atlantic. An excellent swimmer, the goose-sized, upright bird could only waddle awkwardly on land, making it easy prey for European sailors. It was hunted to the brink of extinction for its meat, fat, feathers and eggs. The two birds nesting on Eldey in 1844 were the last great auks to be seen alive.

This human-driven extinction took place in an era when some scholars were still struggling to come to terms with the very idea of extinction in the geological record. The dodo and great auk were easily explained as casualties in a brutal age of global exploration, but the concept of *natural* extinction was hard to accept for those who followed the biblical narrative on life's origins. It violated the widely held belief of the unchanged perfection of Creation. Why on Earth would God go to the trouble of creating an animal and a suitable habitat and then let the creature die out? This would surely upset the balance of nature.

The disappearance of creatures from the fossil record was commonly explained by the erosion of fossil-containing rocks or, rather bizarrely, by arguing that the missing animals were still living and roaming some distant landscape or ocean depths, hitherto unfound. The latter view was commonplace in America into the early part of the nineteenth century where, most famously, the religious convictions of none other than President Thomas Jefferson (1743–1826) made him reluctant to accept the idea of extinction. When bones of the mighty American mastodon (an ice age cousin of the elephant and woolly mammoth) were found at the Big Bone Lick site in Kentucky, Jefferson confidently expected the living beast to be discovered somewhere out in the wild west. This beast so fascinated him that he filled one room of the White House with its bones. But as the century unfolded this position became impossible to sustain when expeditions failed to locate the mastodon or, for that matter, any other creature that had vanished from the geological record. Towards the end of his life even Jefferson accepted that extinction was a possibility.

Georges Cuvier (1769–1832) was the first scholar to provide an objective demonstration that extinction was part of Earth history. Cuvier was remarkably skilled in what became known as comparative anatomy; he has been called the father of palaeontology. By carefully studying the bones of thousands of modern and ancient animals, he identified many species that were no longer living. He assembled a vast collection of fossil bones in the Museum of Natural History in Paris. Cuvier corresponded with Mary Anning and purchased specimens of her Jurassic fossils. In 1800 he examined the remains of a woolly mammoth that had been exhumed from the Siberian permafrost. Cuvier was able to show that it was a cold-adapted species that was anatomically quite different from living elephants. His analysis of the woolly mammoth established extinction as a scientific fact.

By the early nineteenth century major changes in the fossil record were being used to divide up the geological column. In 1841 the English geologist John Phillips (1800–1874), the nephew of William Smith, who produced the first geological map of England in 1815, introduced the three-part scheme of geological eras that map the most momentous changes in the history of life since the start of the Cambrian. Phillips defined the Palaeozoic ('ancient life', the Age of Fishes), Mesozoic ('middle life', the Age of Reptiles) and Cenozoic ('new life', the Age of Mammals) eras, and his scheme is still used today. The boundaries between these eras mark fundamental changes in the history of complex life.

While the Mesozoic is bookended by two major upheavals in Earth history (the Permo–Triassic and the K–Pg extinctions), the full scale of these extinctions would not be recognised for another 130 years. Few things mark the passage of time like a mass extinction, but nineteenth-century natural science certainly was not ready for such catastrophic losses. Twentieth-century geology was slow to catch on too. Why did it take so long for mass extinctions to enter mainstream Earth history?

Charles Lyell is largely responsible for this aversion to catastrophic change. As the most influential geologist of the nineteenth century, he did more than anyone to establish geology as a rational science grounded in field observation and laboratory studies. Building on the ideas of James Hutton from the previous century, Lyell was the most prominent advocate of gradual change in Earth history and argued forcefully that there was no need to invoke catastrophic events to explain the geological record. His legal training came to the fore in his writing – building his case meticulously step by step. Everything in Earth history, he argued, could be explained by observable modern-day processes. This principle is a key pillar of what became known as uniformitarianism (Chapter 6). Lyell's great legacy was the defeat of catastrophism, including the biblical flood. He died in 1875, but his influence endured deep into the next century. Geology became ultra conservative and there are many examples – including Wegener and his drifting continents – of brilliant thinkers who put forward radical ideas only to be branded as cranks.

Lyell and Darwin were close friends and intellectual allies. Both believed that those parts of the fossil record that seemed to show the disappearance of multiple species could be explained by gaps in the rock record. They argued persuasively that large chunks of time were lost; any abruptness in the fossil record could be explained by the many missing pages in the history of life.

Such was the influence of Lyell and Darwin and the conservative instincts of the leading geologists who came after them that mass extinctions did not feature prominently in geology textbooks until the 1980s, when I began my time as a university student. The advent of data analysis by computer in the 1970s and 1980s was also a key step forward as palaeontologists began to compile large databases that brought together fossil records from sites around the world. This new quantitative approach allowed for a much more sophisticated and systematic analysis of the history of life. It played a key role in the recognition of mass extinctions and the quantification of species losses over time. For example, it has allowed palaeontologists to estimate that 99.9 per cent of all species that ever existed are now extinct.

We know now that extinctions happen all the time. The natural background rate of species loss is believed to be about one to five species per year for every million species on Earth. This is the natural baseline for species loss. The fossil record tells us that it is rare for any species of plant or animal to last for more than a few million years. Hundreds of species of dinosaur, for example, including *Brachiosaurus* and *Stegosaurus*, died out in the Jurassic Period tens of millions of years before the asteroid impact at the end of the Cretaceous.

In any ecosystem, some species will be thriving while others are in decline. For some living things, a small shift in the environment, such as a fall in temperature, will accelerate the journey to extinction. If a preferred habitat changes more quickly than a species can adapt, reproductive fitness may be compromised and population decline sets in. The pathway to extinction is set if the animal or plant is unable to migrate to a more favourable environment. The severity and speed of the environmental change will influence whether the extinction is local, regional or global, rapid or slow.

The extinction baseline has fluctuated throughout Earth history but at five periods in the past 550 million years the rate has soared far above this background. These spikes, when most species disappear in a geologically brief period, are called mass extinctions. We have examined these events in earlier chapters:

| | |
|---|---|
| End Ordovician (443.1 million years ago) | ~85 per cent loss (Chapter 16) |
| Late Devonian (360 million years ago) | ~75 per cent loss (Chapter 18) |
| End Permian (251.9 million years ago) | ~95 per cent loss (Chapter 20) |
| End Triassic (201.4 million years ago) | ~80 per cent loss (Chapter 21) |
| End Cretaceous (66 million years ago) | ~75 per cent loss (Chapter 23) |

Earth history tells us that our world has witnessed a series of catastrophic losses when environmental changes conspired to threaten the habitability of our planet. Given these traumas, asking how life has survived for so long may be a more interesting question than asking how life emerged in the first place.

Mass extinctions and the environmental crises that drive them have attracted a great deal of attention – asteroid impacts and belching volcanic landscapes get all the headlines. There has been rather less focus on how to survive such a catastrophe and how animal and plant communities are rebuilt after a mass extinction. The ecological recovery following each mass dying was quite different in terms of the speed of recovery and how the new communities evolved. In the case of the Permo–Triassic extinction, when all life was nearly obliterated, the latest research on the recovery has yielded surprising findings. The marine ecosystem was rebuilt from the top down: the first organisms to emerge in the oceans in the early Triassic were big animals at the top of the food chain, including ichthyosaurs, along with other high-level predators including cephalopods, conodonts (jawless vertebrates) and fishes. (Of course, these high-level predators must have had animals to prey upon – the comparative unavailability of fossils from the lower-level species may partly explain this conundrum in the fossil record.) All these groups diversified in the first few million years after the extinction. The fossil record suggests that the sea floor ecosystem took far longer – about 50 million years – to reach pre-extinction levels of diversity. Studying the fossil record in this way is important because it stresses the value of monitoring and protecting today's ecosystem diversity. Earth history shows us that once this diversity is dismantled in the oceans, the time needed for restoration is much longer – some tens of millions of years.

The K–Pg extinction that wiped out the dinosaurs was a blindingly sudden removal of three-quarters of life on Earth. It is the rapidity of the species loss that sets it apart from the other big extinctions of the last 550 million years. The killing may have been over in less than a decade, while earlier mass extinctions were played out over millions of years.

The ecological consequences of the sudden disappearance of the dinosaurs were profound. The removal of the big sauropod herbivores was a huge change in global ecology, so it is not surprising that we see a major flourishing of land plants early in the Palaeogene in a landscape freed of plant-eating giants. The concentration of fossil pollen preserved in the early Palaeogene rocks increases rapidly at many sites after the extinction event.

The extinction of the dinosaurs has generated many thousands of books and academic papers, yet much less attention has been given to the survivors and the characteristics that helped them to navigate the catastrophe and exploit the niches left behind. This is now an active area of research. Many reptiles such as crocodiles, turtles, snakes and lizards survived the crisis. Various amphibians, including frogs and salamanders, survived into the Cenozoic Era too. Large slices of luck were involved, but we can identify traits that help species survive a mass extinction.

Being small with a large geographical range was a big advantage, as was being able to reproduce rapidly and produce large numbers of offspring. Burrowing was a good strategy too. Most of the mammals that survived into the Cenozoic were rat-sized. Interestingly, amphibious animals also fared rather well. Crocodiles, turtles, frogs, snakes and other creatures that could spend time below water seem to have been sheltered from the climate crisis that followed the asteroid strike. It also pays to be an omnivore when resources are scarce – if you want to navigate a mass extinction, do not be fussy about what you eat. The mammals did suffer losses, but the major lineages, including our ancestors, pulled through.

# CHAPTER 25

# Alligators in the Arctic

Extinctions create opportunities. Every land-dwelling animal weighing more than about 5 kilograms had been eliminated by the environmental crisis at the end of the Cretaceous. The oceans were emptied of big predators. The skies were quiet. Biodiversity had crashed but Cenozoic Earth was primed for a fresh start with a quite different collection of plants and animals.

The Cenozoic Era of 'new life' began 66 million years ago and extends to the present day. It comprises the last three periods in our Earth history: the Palaeogene (66 to 23 million years ago), Neogene (23 to 2.58 million years ago) and Quaternary (2.58 million years ago to present). ('Tertiary Period' is still sometimes used to describe the time covered by the Palaeogene and Neogene, but this term has largely fallen out of favour.) The Palaeogene Period is made up of the Palaeocene, Eocene and Oligocene epochs; the first two are the focus of this chapter. The Palaeocene spans the first 10 million years (66 to 56 million years ago) after the Cretaceous extinction, when Earth's ecosystems began to be refashioned to refill the empty spaces left by the dinosaurs and marine reptiles. The Eocene is just over twice as long, stretching from 56 to 33.9 million years ago. At the boundary of these two epochs a remarkable episode of rapid climate warming coincides with the emergence of the main groups of modern mammals.

As we move closer to the present day, we can generally read more of the geological record: there are fewer gaps due to erosion, and fossils are often better preserved and more easily accessible. We also commonly have a greater range of evidence to be able to build up a picture of past environments and past climates. The soft muds and microfossils that accumulate on the ocean floor, for example, can provide an unbroken high-resolution record of environmental change for the *entire* Cenozoic Era. Using specially modified ships with large drill rigs, we can collect long cores of sediment from beneath the floor of deep-ocean basins and so extract a wealth of physical, chemical and biological evidence about the past.

What kind of Earth did the mammals inherit from the dinosaurs? Once global climate had recovered from the trauma of the asteroid impact, the Palaeocene continued in similar vein to the Cretaceous, with strong greenhouse warming producing an average planetary temperature of about 24 to 25°C – nearly 10 degrees hotter than today's average. Much of the globe was warm and humid. Sea level was about 100 metres higher than today, with an abundance of shallow tropical seas fringing the continents. Even the poles were temperate: Earth was largely ice-free with only tiny glaciers in the highest mountains in the very high latitudes. The Arctic and Antarctic were landscapes vastly different from today, with lush forests and swamps. Greenland really was green.

After a few million years, much of the Palaeocene landscape was covered in dense forest, although there weren't many tree species – the biosphere was still slowly recovering from the ravages of the end-Cretaceous catastrophe. In the polar regions there was extensive broad-leaved deciduous and evergreen woodland and shrubs that had evolved to cope with long winter months without sunlight. There was also beauty. Flowering plants (botanists call these angiosperms) had first appeared in the Cretaceous but during the Palaeocene they steadily increased in diversity in co-evolution with insects who fed on them and aided pollination. The expansion of flowering plants in the warm and wet climate allowed insects to diversify and colonise all the major landmasses. Bees and stinging wasps were mass-extinction survivors that became abundant in this new world of forests, wetlands and flowers. A consequence of a mass extinction is that the average size of animals tends to fall – this has been called the Lilliput Effect. The largest animals in the mid-Palaeocene world were about the size of a Golden Retriever. But they would soon get bigger.

We think of the Arctic as a barren frozen-tundra landscape, but vast tracts were sub-tropical for much of the Palaeocene and Eocene. Greenland was fringed with lush swampy floodplains and its volcanic rocks weathered to form a red soil similar to that which forms in today's humid tropics. Where today we might see polar bears and Arctic foxes, the landscape was inhabited by cold-blooded reptiles and small hippo-like beasts. Alligators and turtles wallowed in forest-lined wetlands.

As in the Carboniferous, the Palaeocene swamps were compressed into coal deposits over millions of years. In Svalbard, which was situated close to the northern part of Greenland in the Palaeocene, two miners discovered well-preserved fossil footprints of *Titanoides* exposed in the rocks forming the roof of a mine shaft in 2006. This beast was part of a group of mammals called pantodonts that went extinct towards the end of the Eocene. *Titanoides* was a stocky herbivore with a bear-like appearance, up to 3 metres in length and weighing about 150 kilos. The tracks in the coal beds represent the earliest evidence of a large mammal on the Arctic islands and the northernmost record from the Palaeocene. *Titanoides* migrated to Svalbard from North America.

At the same latitude in Arctic Canada, the fossil record on Ellesmere Island shows that alligators, giant tortoises and aquatic turtles thrived here some 52 to 53 million years ago and coped with six months of darkness. Temperatures never fell below freezing on Ellesmere Island in the early Eocene, while average temperatures

for the warmest month were about 19 to 20°C. In the Palaeocene–
Eocene greenhouse, the climate of Arctic Canada was more like
Florida today. Ellesmere Island hosted a warm temperate swamp-
forest ecosystem like today's cypress swamps of the southeastern
United States.

A single toe bone of *Gastornis* has been found on the island. The
Palaeocene and Eocene forests saw the emergence of giant flightless
birds such as *Gastornis*, with massive skulls and terrifyingly huge
beaks, which represent an unbroken line to the dinosaurs. They
were not-so-distant cousins of *Velociraptor*, although *Gastornis* was
vegetarian, using its enormous beak to tear at foliage, seeds and
hard fruit. An extinct family of carnivorous, fast-running 'terror
birds' emerged in South America in the middle Eocene – the
phorusrhacids were apex predators who used violent kicking to
overcome their prey before ripping their flesh with the sharp
curved tips of their beaks.

In the case of seabirds, a key evolutionary development in the
early Palaeocene was the transition from flight to diving: we see the
earliest penguins in this period. Fossil bones of giant penguins
have been discovered in New Zealand and Antarctica, where one of
these ancient penguin species stood about 1.65 metres tall and was
four times heavier than today's emperor penguin. These early
penguins were able to evolve to such a size because the large flesh-
eating marine reptiles that dominated the ocean for over 100
million years were eliminated at the end of the Cretaceous. This
left the oceans empty of big predators.

During the Palaeocene and Eocene, Earth's climate was punctuated by a series of abrupt jumps in temperature called hyperthermals. These were short-lived episodes of rapid greenhouse warming,
ranging from about 10,000 to 200,000 years' duration, but long
enough to trigger widespread impacts on landscape processes,
ecosystems and the distribution of plants and animals. Their effects
included ocean acidification, increases in storms and big floods,
habitat shifts and widespread extinctions.

The hyperthermals were triggered by massive injections of carbon
dioxide into the atmosphere and oceans. The most prominent

hyperthermal took place at the close of the Palaeocene Epoch, between 55 and 56 million years ago. It is known as the Palaeocene–Eocene Thermal Maximum (PETM) and saw rapid global warming of between 5 and 9°C on land and in the oceans over a period of less than 200,000 years. Sea-surface temperatures increased by up to 9°C and were higher than at any time in the past 66 million years.

The warming associated with the PETM was geologically rapid, with the spike in temperature taking place within less than 10,000 years. This super-hothouse Earth saw rapid changes in ecology on land and in the oceans. The atmosphere of the PETM would have held more moisture than today, so powerful storms were more frequent. The rock record in the Spanish Pyrenees at the PETM gives evidence for exceptionally large river-rolled boulders fanning across the landscape with river channels much bigger than those of today. These features point to the work of intense storms and large floods under a monsoonal climate.

The PETM affected all parts of the Earth system, but its impacts are imprinted especially vividly in the marine sediment record. The ocean is a major carbon sink. When carbon dioxide builds up in the atmosphere, the oceans are forced to absorb more of this gas. Ocean water then becomes more acidic and the chemistry of the marine environment changes. This can be catastrophic for corals and for all creatures that build shells because the concentration of carbonate ions in the water decreases. After absorbing vast amounts of carbon dioxide 56 million years ago, the oceans suddenly became more acidic.

The deep ocean floor normally comprises a soft grey-white ooze made up of billions of tiny calcium carbonate shells of dead single-celled marine organisms called foraminifera (forams) that have sunk from their living habitats in the water column above. These ocean-floor deposits can be thousands of metres thick. We can see a sharp colour change from white to red-brown clays marking the PETM, because carbonate shells were unable to form in this warm acidified ocean. This distinctive layer was first identified in the 1980s when long sediment cores were collected from the floor of the Southern Ocean near Antarctica.

The community of organisms that live at the bottom of an ocean, lake or river is known as the *benthos* (from the Greek meaning 'deep of the sea'). Benthic forams in the deep ocean suffered a global extinction during the PETM, with about 50 per cent of species lost. These foram species had survived the catastrophe of the end-Cretaceous mass extinction but were wiped out by the rapid warming at the PETM.

The origin of the greenhouse gases that powered the warming has been much debated. As the continents of North America and Europe moved apart in the Palaeocene and Eocene there was widespread volcanic activity in the North Atlantic region that produced huge volumes of carbon dioxide. But these inputs may not have happened quickly enough to account for the pronounced warming spike. A more powerful greenhouse gas may have played a role. A compound called methane hydrate is stored in shallow sediments on the ocean floor. It is produced from rotting organic material in buried sediments. The methane gas migrates upwards and becomes frozen as methane hydrate just below the seabed close to the very cold ocean-bottom waters. If ocean temperatures increase rapidly, enormous quantities of methane hydrate can thaw and be released into the atmosphere. These 'methane burps' can rapidly warm the climate because of their potent greenhouse properties. This is an attractive explanation for the rise and fall in temperature associated with the PETM spike because once the seabed methane supplies are exhausted the climate would quickly shift to a cooler mode.

The PETM has become a hot topic because it is a remarkable interval of rapid warming that saw widespread biogeographical and evolutionary change, with gains and losses across the animal kingdom. New groups of mammals appeared and dispersed across all continents, with new routes opened in the high latitudes. Australia and South America were still connected to Antarctica at this time as part of the Gondwana landmass.

Key elements of today's biodiversity were put in place during the PETM. It coincides with the emergence of the three main orders of modern placental mammals across North America, Asia and

Europe. Two of these orders are large mammals with hooves known as *ungulates*: we see the even-toed ungulates (artiodactyla) that today include species of deer, giraffes, sheep, goats, pigs, camels, llamas, alpacas, hippopotamuses, antelopes and many others; and the odd-toed ungulates (perissodactyla), which today include species of horses, asses and zebras, as well as the rhinoceroses and tapirs. The primates appeared too – today this group includes species of monkeys, apes, lemurs, bush babies, tarsiers and, of course, humans. There were also many losses of animal species during the PETM. *Champsosaurus* was a crocodile-like reptile with a long snout and powerful jaws that lived in rivers and lakes. It survived the chaos of the end-Cretaceous mass extinction but disappeared from the fossil record during rapid biogeographical changes of the PETM.

The hyperthermals of the Cenozoic are also of great interest to climate scientists because they may help us to better understand what might lie in wait for the Earth's climate in the coming decades. In the second half of the Eocene, about 50 million years ago, Earth started to cool, and temperatures dropped steadily so that it became cold enough for large glaciers to form in Antarctica. Earth had begun to shift from hothouse to icehouse.

# The Last Great Cooling

One hundred and twenty million years ago a huge chunk of land that we now call India broke away from the Antarctic landmass and began a long, slow journey north. Sixty million years later it had drifted to the equator. About 50 million years ago, in the Eocene Epoch, it collided with southern Asia and pushed up what is now the greatest mountain range on Earth. The collision generated immense tectonic forces that also elevated Tibet, the world's largest and highest plateau region, which averages over 4,500 metres above sea level. The Indian subcontinent is still inching northwards today, squeezing and buckling the crust to sustain the dizzying heights of the Himalayan peaks.

In terms of the hothouse climate, polar forests and global continental geography, Palaeocene and Eocene Earth was much like the Cretaceous but without the dinosaurs. But things were about to change. The formation of the Himalayas and the Tibetan Plateau coincided with a big shift in global climate that would profoundly alter our planet. Earth began to cool so that glaciers advanced and

small ice caps formed in the mountains of Antarctica. By the onset
of the Oligocene Epoch, 33.9 million years ago, global climate had
cooled enough to allow a huge ice sheet to develop that overwhelmed the Antarctic continent. The giant southern landmass
had been largely ice-free since the Permian Period some 250
million years previously; now the Eocene forests were buried
under 4 kilometres of ice. The Oligocene Epoch saw the cryosphere
return as a major player in the global climate system. Earth had
entered a new ice age.

The Oligocene lasted for just 11 million years, but it was a time
of important geographical changes when key features began to
emerge that we recognise in today's Earth system. Back in the fifth
century BC, the Greek philosopher Parmenides was the first to
discuss the idea of an Earth with climate zones. He talked of a
frigid north and a torrid southern zone of sun-baked lands on
either side of a more temperate environment. And it is in the
Oligocene Epoch that the great climate belts began to form as the
tropics slowly contracted towards the equator, deserts and grasslands expanded and more seasonal and temperate climates emerged
in the mid-latitudes. The top and bottom of the world became
increasingly cold. As both poles cooled, a greater contrast in
temperature with the tropics emerged. The major ocean currents
that transport most of the heat around the globe were strengthened
considerably as the Earth system tried to even out the differences
in temperature.

We know that thick glacial ice covered large areas of Antarctica
at the beginning of the Oligocene because long sediment cores
extracted from the seabed hundreds of kilometres from the
Antarctic coast store critical evidence. These well-dated cores
contain distinctive layers of fine rocky debris eroded from the interior of the Antarctic continent. This material could only have been
transported such a distance offshore by drifting icebergs that had
calved away from glaciers extending down to sea level. When
glacier ice is found at sea level we are dealing with a big ice sheet.
Geochemical evidence from tiny shells in these ocean sediment
cores suggests a rapid phase of ice build-up at the beginning of the

Oligocene that formed a vast ice sheet in less than 40,000 years. A positive feedback was in place: ice-sheet growth was encouraged by bright snow-covered ice reflecting solar energy back into space; cooling promotes more permanent snow and ice, which in turn promotes further cooling. The Palaeocene and Eocene had seen some of the warmest temperatures of the last 500 million years. Why did Earth's climate shift to icehouse mode?

The global cooling during the transition from the Eocene to the Oligocene Epoch has been described as the most important climate tipping point of the last 65 million years. The main cause of this cooling was a weakening of the Earth's greenhouse effect. American geologist Maureen Raymo (b. 1959) and two of her colleagues, Philip Froelich (b. 1946) and Bill Ruddiman (b. 1943), proposed a theory to explain the cooling which they called the *uplift weathering hypothesis*. This was a controversial idea in the late 1980s, when they first argued that 'as mountains go up Earth's temperature goes down'.

At the onset of the Oligocene, like today, the Himalayas were the most extensive and highest mountain range on Earth and the Tibetan Plateau was the largest area of uplifted crust. This region is often called the Third Pole because it holds more snow and glacial ice than anywhere in the world outside the Arctic and Antarctic. Meltwater from these highlands feeds a network of big rivers, including the Ganges, Brahmaputra, Mekong and Yangtze. Each transports huge volumes of water and sediment to the ocean.

High mountains are hotspots of weathering and erosion. Glaciers grind down the bedrock, rivers carve deep valleys, earthquakes trigger landslides on the steep slopes, freeze-thaw action splits rocks apart and rainwater attacks rock fragments and breaks down the minerals. In the long term, all this erosion and rock disintegration wearing down the landscape counters the relentless tectonic forces heaving the mountains skywards.

The monsoon climate in this part of Asia is intensified by the presence of the Tibetan Plateau, which acts like a giant hot plate warming the atmosphere. Rising air masses lift moisture upwards from the warm ocean, producing very heavy summer rains. As the colossal stacks of Himalayan rocks are broken down by physical and chemical weathering, the carbon dioxide in the rainfall combines with the weathered material to form new compounds, including calcium carbonate ($CaCO_2$). When this material finally reaches the ocean, trillions of plankton use it to build their tiny shells. It is also the raw material for building corals and every shape of shell in the marine environment.

When plankton die in the open ocean their calcium carbonate shells rain down onto the seabed where they are buried, eventually becoming limestone rock. When carbon dioxide is drawn from the atmosphere faster than it is replenished by processes such as volcanic activity, the greenhouse effect is weakened and the global thermostat is set to cool (Chapter 17). Because of their vast scale and the wet monsoon climate, the formation of the Himalayas created a new global hotspot for carbon capture. Maureen Raymo has described the Himalayas as a huge sponge pulling $CO_2$ out of the atmosphere, and believes they are in large part responsible for all the global cooling in the late Eocene and Oligocene.

Plate tectonic movements produced another profound change in the Oligocene Epoch by isolating the Antarctic continent as Australia and South America broke away and drifted northwards. This created the largest flow of water on Earth. The Antarctic circumpolar current (ACC) flows clockwise around Antarctica, connecting all three major ocean basins: Atlantic, Pacific and Indian. It extends from the ocean floor to the sea surface, controlling the movement of heat, carbon and nutrients (including plankton) throughout the Southern Ocean.

So how big is this flow around Antarctica? The Sverdrup (Sv) is the unit used by oceanographers to describe the size of an ocean current. One Sverdrup is a flow of 1 million cubic metres of water per second – five times the flow of the Amazon. The combined flow of all the rivers in the world is about 1.2 Sverdrups. The flow of the ACC through the 1,000-kilometre-wide Drake Passage between the Antarctic Peninsula and the tip of South America is over 150 Sverdrups. Powerful westerly winds drive the surface waters of this flow. Generations of sailors have experienced the treacherous seas where the Drake Passage funnels the narrowest and strongest part of the ACC.

The ACC plays a vital role in Earth's climate system through its influence on global ocean circulation. It is crucial for the habitability of our planet because it helps to keep Antarctica cold and frozen. The flow is so large and so strong it prevents warm currents from the Indian, Pacific and Atlantic oceans from reaching the Antarctic continent. On the outer edge of the ACC, where the cold Antarctic flow meets warmer waters, there is a great upwelling of nutrients that support billions of plankton and krill. This zone of high productivity provides rich pickings for many species of fish, squid, seals, penguins and whales, as well as the great albatrosses of the Southern Ocean (Chapter 29).

As the Oligocene climate became cooler, large areas of the globe became drier, tropical wetland forests gave way to more temperate woodlands, and trees gave way to grassland ecosystems that would become even more extensive in the following Miocene Epoch. This was the emergence of the grand global pattern of biomes.

Today we recognise several biomes, including tropical rainforests, savannah grasslands and hot deserts. In the cold high latitudes of the northern hemisphere we find the boreal forest biome; the vast region of coniferous woodland stretching across much of Russia and Canada is the largest land biome. Beyond these forests lies the Arctic tundra. This global pattern of biomes is familiar to us – we have seen the coloured zones mapped out in geography textbooks and on classroom walls. In geological terms they are quite a recent development in Earth history that reflects the changing geography of the continents and the evolution of today's global climate system.

The evolution and expansion of grasslands – the North American prairies, the steppes of central Asia and the savannah of Africa – is one of the most significant changes of the Cenozoic Era. The grassland biome co-evolved with many species of grazing animals such as horses and camels; these and other hooved animals increased in size during the Oligocene. The first elephants with trunks and the first primates appeared in Africa.

The remarkable transition from the Eocene hothouse to the Oligocene icehouse saw a series of extinctions on land and in the oceans. None were mass extinctions like the big five we have already explored, but the stresses of this period saw a major turnover of animal species. Many ancient mammal species (including all tree-dwelling mammals) went extinct in Europe and were abruptly replaced by species from Asia including rabbits, ancestral species of rhino, and a family of animals called the *Anthracotheriidae* that looked rather like underfed hippos with small heads. This rapid replacement may have been enabled by the lowered sea level that opened migration corridors from Asia.

While the imperfect fossil record indicates that the overall diversity of mammals was rather low in the Oligocene, there were some extraordinary beasts, including the largest land mammals that ever lived. The paraceratheriids were a family of long-necked, long-legged hornless rhinos that lived in Asia and Eastern Europe. Their skulls and legs were longer than those of all known land mammals. The largest adults stood almost 5 metres at the shoulder and could attain a body mass of 20 tonnes – equivalent to four African elephants. Their long necks and teeth suggest they were browsers, feeding on leaves and woody plants. Complete skeletons of these giant rhinos are quite rare so there was initially some confusion about their body plan. Recent discoveries in China and on the Tibetan Plateau have improved our understanding of these remarkable animals. The giant rhinos experienced an unusually rapid increase in body size in the late Eocene as the climate cooled and more open woodland and grassland habitats emerged. When sea levels fell in the Oligocene these giant rhinos expanded their range along the shores of the ancient Tethys Ocean, which extended along the southern coastline of Asia.

As more glacial ice became locked up on the Antarctic continent, sea levels fell, forcing extinctions in shallow marine habitats that were replaced by wide coastal plains. For much of the Eocene, the Arctic Ocean had been effectively isolated, surrounded by land, but as the seaway between Greenland and Norway opened it became connected to the global ocean. In the Oligocene it became a source of frigid bottom waters that chilled habitats in the deep ocean so that many marine species were wiped out. The cooling of the poles profoundly affected life in the oceans, where habitats became more fragmented. Creatures that could live in cooler waters became dominant at higher latitudes.

While the Oligocene is a geologically brief epoch, it saw major changes in global geography, climate and vegetation. It was a time of transition in Earth history when new biomes emerged as the climate became cooler and drier. These changes were driven by large-scale tectonic processes that pushed up the Himalayas and isolated Antarctica. In the Miocene Epoch that followed, tectonic forces triggered an ecological catastrophe in the Mediterranean Sea.

# CHAPTER 27

# Crisis in the Mediterranean

The Mediterranean is a warm, salty sea. Since ancient times, civilisations have harvested salt there by evaporating seawater under the scorching summer sun. The Romans built salt pans at the mouth of the River Tiber and Venetian merchants acquired great wealth in the Middle Ages exporting salt produced in the lagoon. You can see working salt pans across the Mediterranean today. At Marsala on the westernmost tip of Sicily, a palette of watery salt-pan squares grades from blue-green to salt-heavy deep pink. At the end of the production line are cones of brilliant white.

In 1970 a team of marine geologists aboard the *Glomar Challenger* research vessel began drilling into the floor of the Mediterranean, seeking clues about the deep history of the basin. In the upper few hundred metres of marine sediment they found just what they expected – stiff muds, thin layers of dark organic matter and the creamy ooze of plankton microfossils. Below these sediments they drilled into layers of salt more than 1,000 metres thick. Everywhere they drilled they found hundreds of metres of salt. The core samples they hauled up from below the seabed provided the first definitive evidence that the deep blue waters of the Mediterranean had almost completely dried out in the recent geological past.

This episode is known as the Messinian Salinity Crisis (MSC), after the city of Messina in northeast Sicily where these salt deposits are prominently exposed after being raised from the seabed by tectonic uplift. The MSC took place at the very end of the Miocene Epoch. It has been described as the most abrupt, global-scale environmental shock since the asteroid impact that annihilated the dinosaurs at the end of the Cretaceous.

The Mediterranean lies next to the Sahara, the largest hot desert on Earth. In a typical year, the sea loses almost a metre of water to evaporation. Since regional rainfall and the rivers that flow into the sea do not supply enough water to replenish this loss, the annual water balance of the Mediterranean basin shows a big deficit. Even the mighty Nile flood has only a negligible impact on this shortfall. The difference is made up by the inflow of water from the Atlantic Ocean through the Strait of Gibraltar – a connection with the global ocean that is of critical importance to maintain sea level in the basin. Without this Atlantic inflow the Mediterranean could be emptied by evaporation in a geological instant of around a thousand years. This fact alone tells us this is a precarious marine environment where conditions could change very rapidly.

There are two main flows through the Strait of Gibraltar: one enters and one leaves the Mediterranean. Cool water from the Atlantic flows eastwards into the basin at the surface because it contains less salt and is therefore less dense. Strong evaporation in the Mediterranean, especially in the eastern basin beyond Sicily, creates water with a higher salt content which is therefore denser. This deeper, saltier layer flows in the opposite direction through the Gibraltar Strait into the Atlantic Ocean. Remarkably, this deep saline current can be traced northwards around the Iberian Peninsula as well as across the Atlantic all the way to the Caribbean. In the Second World War, German U-boats could ride these powerful currents in and out of the basin with their engines shut down to avoid detection. After accounting for the deeper outflow, the net inflow from the Atlantic of roughly 2,000 cubic kilometres per year is critical to offset the Mediterranean's water deficit. Without this inflow the Mediterranean would steadily lose volume and become saltier. The inflow of Atlantic water is also important for the ecological health of the marine environment because it is one of the primary sources of nutrients for the Mediterranean.

Towards the end of the Miocene Epoch the water balance was disrupted as the African plate rumbled gradually north and interacted with the much smaller Alboran plate that today sits between Spain, Morocco and Algeria. The African plate squeezed the landscapes and marine connections at the western end of the Mediterranean, narrowing the Atlantic connections and making them shallower. (It continues to do so: today the bedrock sill in the Strait of Gibraltar sits just 300 metres below the sea surface. At its narrowest the continents of Africa and Europe are less than 15 kilometres apart.) Atlantic waters were still able to enter the Mediterranean, but the deeper, salt-rich outflow became blocked. We do not know if the connection ever closed completely, but this scenario allows for salty water to flow in only one direction, so the Mediterranean became increasingly saline. As the level of the Mediterranean plummeted by up to 2 kilometres, it rapidly transformed into a series of barren briny basins rather like today's Dead Sea.

When the inflow from the Atlantic was severely restricted, the rivers Nile, Ebro, Rhone and Po – as well as several rivers in North Africa that no longer flow today – would have influenced the chemistry of the saline basins they flowed into. Depending on local conditions of temperature, chemistry and water depth, great thicknesses – about 1.5 kilometres on average – of halite (common salt) and gypsum were deposited. These salts are known as evaporites. Geophysical surveys have located the upper and lower boundary of the Messinian salt deposits across the Mediterranean. The upper surface forms a distinctive and broadly horizontal layer across much of the basin – a colossal slab of salt from Spain to Lebanon. These evaporite deposits include one of the largest bodies of halite on Earth.

The latest dating shows that the Mediterranean became effectively landlocked towards the end of the Miocene between about 5.97 and 5.33 million years ago. This is a *very* brief geological window to account for the huge thickness of evaporite deposits and the refilling of the sea – all of this happened remarkably quickly. The total volume of salt deposited in the basin during the MSC could be more than 2 million cubic kilometres. The Mediterranean's massive accumulations of evaporite deposits are called salt giants and they have formed in other periods in Earth history.

During the salinity crisis, the Mediterranean became the most effective salt pan on the planet. Since almost all this salt was taken from the global ocean and trapped in long-term storage on the floor of the Mediterranean, the world's oceans became less saline in the late Miocene, with 6 to 10 per cent of the total dissolved salt load in the global ocean dumped in the Mediterranean. A minor tectonic uplift at the western end of the Mediterranean triggered a chain of events that changed the composition of the global ocean.

Today the shoreline of the Dead Sea is the lowest land on Earth, sitting some 440 metres below global sea level. It stretches our geological imagination to consider that the water level in the Mediterranean basin during the MSC was at least several hundred metres lower than the surface of the global ocean; at times it may have been 2 kilometres lower. As sea level fell, the rivers around the Mediterranean were forced to cut down into bedrock as they flowed over great cataracts into a rapidly emptying basin. Most spectacularly, the River Nile carved a late Miocene bedrock canyon 1,000 kilometres long all the way upstream to what is now Aswan. This ancient Nile Canyon, now buried beneath floodplain and desert, is longer, deeper and wider than America's Grand Canyon.

There is a large body of evidence to suggest that the refilling of the Mediterranean was accomplished by the most catastrophic flood event recorded in Earth history. This is known as the Zanclean megaflood. It was responsible for the rebirth of the Mediterranean Sea that we know today. The Zanclean (5.33 to 3.60 million years ago) is the first stage of the Pliocene Epoch that follows the Miocene. All across the Mediterranean, the onset of the

Zanclean coincides with the end of the salinity crisis when the basin was refilled.

The reconnection with the Atlantic may have been caused by faulting and erosion at the bedrock sill in the Strait of Gibraltar or by sea-level rise following partial melting of the Antarctic ice sheet. Whatever the initial cause, the flood that refilled the Mediterranean would have started as a small trickle in the Strait and grown into a mighty river – fed by a waterfall from the Atlantic – that began to cut down into its bed. As the flows increased, the river-cut notch deepened to produce deeper and wider erosion so that, fed by the global ocean, the floodwaters became colossal. At its peak, the Zanclean flood may have carried the volume of a thousand Amazons through the Strait. The dried-out Mediterranean was refilled in a geological instant, perhaps in less than two years. Sea level would have risen by a staggering 10 metres per day. Reconstructions of this rapid refilling show peak floodwaters thundering through a bedrock canyon south of Sicily at over 100 kilometres per hour, cascading down a waterfall 1,500 metres high! Other seas have dried out in Earth history, but this was surely the most dramatic rebirth.

The sediment cores collected by the *Glomar Challenger* show an abrupt shift from the salt deposits of the MSC to the overlying marine sediments characteristic of open-sea conditions, suggesting a rapid transition from shallow briny basins to a refilled sea. The fossil record also supports a rapid refilling in the Zanclean because there is a sudden reappearance immediately above the salt deposits of many species of marine creatures that could only survive in open-sea conditions. The sea became very deep very quickly.

A key piece of evidence for a megaflood comes from the topography of the sea floor in the western Mediterranean, where mapping has revealed a huge canyon almost 390 kilometres long and several hundred metres deep, stretching from the Gulf of Cádiz (on the Atlantic side) to a deep basin off the coast of Algeria. Atlantic water must have flooded in with spectacular violence to cut such an immense feature in bedrock.

If the catastrophic-refilling model is to hold up, the huge channels that were eroded in the western Mediterranean must have produced equally voluminous flood *deposits* further downstream, since the eroded sediments had to go somewhere. Surveys of post-Messinian deposits deep below the seabed in the Ionian Sea between Italy and Greece have identified an extensive body of highly chaotic sediments lying above the evaporites. This is consistent with the passage of a huge flood from the western to the eastern Mediterranean basin via a route to the southeast of Sicily. The deposits are a jumbled mix of pebbles and boulders almost 800 metres in thickness, suggesting they were deposited rapidly and haphazardly by a catastrophic flood event. This body of sediments has been described as the largest known megaflood deposit on Earth.

While the geological and environmental impacts of the MSC have been studied intensively for decades, the *ecological* crisis in the saline marine basins has only recently been systematically examined. Detailed analysis of the Miocene and early Pliocene fossil records for the Mediterranean – from tiny plankton to big marine predators – has provided fascinating insights into the decline and recovery of a major ecosystem subjected to desiccation.

The eastern end of the Mediterranean was closed off in the early Miocene. It was previously connected to the Indian Ocean and part of a major ancient tropical ocean called Tethys that extended along much of the southern shore of Asia. For the rest of the Miocene, the Mediterranean was an inland sea with a sole connection to the global ocean via the narrow western gateway. This relative isolation saw the evolution of many endemic species in the Mediterranean, but most were lost during the MSC. Unsurprisingly, the crisis was catastrophic for marine life, with reef-building corals and most of the fish species who inhabited them disappearing completely from the Mediterranean. The fossil record shows almost eight hundred species of marine creatures endemic to the Mediterranean before the MSC. Only eighty-six of these species (11 per cent) survived the crisis. Biodiversity in the Mediterranean took almost 2 million years to recover, and it was different than it had been before: the fossil record tells us that before the salinity crisis biodiversity was highest in the far eastern part of the Mediterranean – a legacy of the ancient connection with the Indian Ocean – but after the refilling, it has been, and continues to be, higher in the western basin. But it could not replace what was lost.

When the waters of the Atlantic began to flow back into the Mediterranean, they brought life back into the basin. Some 2,700 species hitched a ride on the floodwaters and repopulated the sea. Some of these species had inhabited the Mediterranean before the great desiccation, but most were first-timers – species that had never previously inhabited the Mediterranean – including species of tuna, swordfish, jellyfish, dolphins and great white sharks. Most of the species from the old Mediterranean were lost as the hyper-saline waters became uninhabitable for everything but a few hardy salt-tolerant species. Some fish and bivalves found refuge in river estuaries, but almost everything else died out. This was the Mediterranean's mass extinction.

# CHAPTER 28

# Joining the Americas

On the southwestern tip of Scotland there is a botanic garden stuffed with plants from distant lands. There are tree ferns from New Zealand, eucalyptus groves, and a species of rowan from Madeira that is one of the rarest trees in the world. This subtropical oasis is possible because 3 million years ago, plate tectonic processes joined South America with North America and created one of the strongest ocean currents on Earth. The Logan Botanic Garden sits on a narrow peninsula at the top of the Irish Sea that is bathed in warmth from the Gulf Stream.

For much of the Cenozoic Era, South America and North America were island continents separated by a wide body of water known as the Central American Seaway. This gap allowed warm tropical water to move freely between the Atlantic and Pacific Oceans. In the Miocene Epoch, as the Cocos tectonic plate slipped beneath the Caribbean, plumes of magma rose upwards, forming pillow lava and a cluster of seabed volcanoes. By about 15 million years ago, an arc of volcanic islands emerged between the continents that began to slow the exchange of water. As the plates jostled, the ocean floor bulged, leaving the seaway shallow enough for rivers to fill the gaps between the islands. Layers of sediment were stacked on the sea floor until a continuous strip of land joined the continents together.

It is difficult to be precise about the timing of the final closure, but by 3 million years ago, towards the end of the Pliocene Epoch, the Isthmus of Panama was in place – North and South America were connected for the first time since the breakup of the supercontinent Pangaea. Isthmus comes from the Greek *isthmos*, meaning 'neck'. When you look at an atlas of the world, South America appears to be dangling precariously by the thinnest of threads. The creation of this strip of land is one of the most momentous events in recent Earth history. Not only did it modify global ocean circulation and play a role in the cooling of the Arctic, it also triggered animal and plant migrations, which caused profound ecological transformations in both North and South America.

In 1513, when the Spanish conquistador Juan Ponce de León (*c.* 1474–1521) led the first European expedition to Florida, his ships met a current so strong it drove them backwards. This was the first documented human encounter with the Gulf Stream. In a flow of giant eddies and looping meanders, this warm, salty current conveys heat polewards from the tropical Atlantic to the Greenland and Norwegian seas. Powered by strong trade winds and differences in water density, the Gulf Stream tracks the eastern coastline of the United States as far as North Carolina, where the continental shelf drops sharply into deeper ocean waters. Here the current pivots slightly eastwards, traversing the Atlantic towards Europe. The Gulf Stream transports a colossal amount of energy – equivalent to that generated by a million nuclear power stations. Its northern extension, the North Atlantic Drift, prevents sea ice from forming along much of the coast of Norway. Once the Central American Seaway was plugged and ocean currents rerouted, the Gulf Stream became much stronger as flows that had previously entered the Pacific were deflected northwards, delivering tropical warmth to western Europe.

The Gulf Stream is a powerful surface current – a river within an ocean creating warm air masses that make the climate milder wherever they pass over land. Bermudans can sip cocktails on the beach while, on the same latitude, the coast of North Carolina endures freezing winter storms. In Europe, the warming influence of the Gulf Stream fades as you move deeper into the continent and winters become much more severe. Consider the blisteringly cold winters and iron-hard ground of Kyiv in Ukraine at the same latitude as Cornwall in England, where winters are typically mild and frost free. Without the warmth flowing from the Gulf of Mexico, winters on the Atlantic fringe of northwest Europe would be colder by some 5 degrees Celsius.

The Isthmus of Panama was completed during the Pliocene Epoch, which spans the period from 5.33 to 2.58 million years ago. Together the Miocene and Pliocene form the Neogene Period – last but one in the geological timescale. For most of the Pliocene, Earth's climate was warmer than today and sea level was several metres higher because there was less water on the continents stored as glacial ice. Between 4.4 and 4.0 million years ago, mean global temperatures were some 4°C higher than pre-industrial values – an interval known as the Pliocene Climatic Optimum. It saw the warmest conditions of the last 5 million years. This warm period is of particular interest to climate scientists because the concentration of carbon dioxide in the Pliocene atmosphere at that time ranged from 400 to 450 parts per million. Today's value (423 ppm as I write this chapter) falls slap bang in the middle.

The Pliocene was a momentous time of global change, when key features of today's global physical geography were put in place. The latter part of this epoch saw global cooling and the formation of extensive and permanent sea ice in the Arctic Ocean, as well as ice caps and glaciers in Greenland and the Arctic islands. Greenland was the first landmass in the northern hemisphere to see the formation of a continental-scale ice sheet, and has supported a big ice sheet ever since, although it has waxed and waned throughout its history.

The conditions that enabled the onset of large-scale glaciation in the Arctic have been called the late Pliocene climate crash. The top of the world became frozen – perhaps for the first time since the Snowball Earth episodes some 600,000 million years earlier. Indeed, one of the big puzzles of the Cenozoic Era is why it took so long for the Arctic to become cold enough to freeze up, since the Antarctic had developed a big ice sheet some 30 million years earlier at the beginning of the Oligocene. The Arctic may have been warmer in the earlier Pliocene because warm Pacific water entered via the Bering Strait between Alaska and northeast Russia.

Before we look at the cooling of the Arctic it is important to appreciate that the physical geography of Earth's polar regions is vastly different. Both poles are capped by ice, but ice with a quite different origin that behaves in very different ways.

The Antarctic is a huge continent sitting over the South Pole, surrounded by an ocean. It is dominated by *glacial* ice that has built up on land fed by snowfall. In places the Antarctic ice sheet is over 4.5 kilometres thick. At the other end of the Earth the Arctic is its polar opposite. There is an *ocean* at the North Pole, almost completely surrounded by land – the vast coastlines of Russia, Canada and Alaska. So the ice at the North Pole is sea ice, *not* glacial ice. Sea ice is frozen seawater that can attain a thickness of a few metres. The Arctic region also has many glaciers and ice caps and a huge ice sheet in Greenland, but the top of the world is an ocean with a lid of sea ice that changes in extent from summer to winter.

Sea ice was present in parts of the Arctic in the Miocene, but it was patchy and limited in extent. In the late Pliocene it became a permanent feature. Sea ice that survives the summer melt season is known as multi-year ice. The reasons for the onset of extensive sea ice formation and land-based glaciation in the Arctic towards the end of the Pliocene (for which we have good evidence from marine sediment cores) have been strongly debated, but many geoscientists argue that the closing of the Central American Seaway set the stage for a widespread freeze and glacier development at the top of the world after 2.7 million years ago. This has been called the *Panama Hypothesis*. However, it does seem somewhat counterintuitive. How might such a major poleward flow of heat freeze up the high latitudes? One might expect it to have the opposite effect.

One explanation is that the enormous amounts of moisture evaporated from the Gulf Stream and North Atlantic Drift fell as rain and snow over Europe and northern Russia. This would send more fresh water via rivers flowing into the Arctic Ocean (making it less salty) so that sea ice could form more easily. This is where feedbacks kick in, especially if the region was close to a tipping point and only needed a nudge to shift to a new climate regime. Sea ice reflects solar energy back into space and a lid of ice on the Arctic Ocean keeps the atmosphere cold because any heat in the ocean cannot escape. The cooling of the Arctic encouraged the formation of glaciers on land. Once sea ice, snow cover and glaciers became extensive, they cooled the regional climate even further.

There has been much debate about how important the closure of the Central American Seaway was as a trigger for initiating cooling and widespread glaciation in the high latitudes of the northern hemisphere. Some modelling studies support the theory while others do not. There is little doubt that the narrow strip between the Americas intensified the Gulf Stream and changed the global climate system. The increased moisture supply to the North Atlantic was a key step on the path to glaciation. But the final trigger may have been a slight change in the amount of solar radiation received in the far north, following a shift in the tilt of the Earth around 2.8 million years ago. This made summers cooler and allowed more snow and ice to survive year-round.

As the Arctic was getting colder in the later Pliocene and tundra ecosystems were developing in the far north, big ecological changes were taking place either side of the new Isthmus of Panama. After the extinction of the dinosaurs, North and South America evolved distinct mammal communities because they had been geographically separated for most of the Cenozoic.

The Great American Biotic Interchange is perhaps the best-documented large-scale faunal interchange in Earth history. The fossil record shows that the movement of animals across the isthmus took place in a series of pulses over the 1.5 million years after the closure. North American species like felines, canids, bears, deer, camels and rodents moved into South America. Many of these species thrived, leading to significant ecological changes and extinctions among native South American fauna. The llama is one of the iconic animals of South America. Yet, along with its cousins the alpaca and the two wild South American camelids, the guanaco and vicuña, the llama is only in South America today because of the land bridge provided by the Isthmus of Panama. Their now-extinct ancestors evolved on the plains of North America in the Oligocene and Miocene before migrating south from the northern continent during the Great American Biotic Interchange. Conversely, South American species such as armadillos, opossums, ground sloths and porcupines moved into North America, though fewer of these species established viable long-term populations. The South American terror birds (phorusrhacids), for example, failed to establish a lasting presence in North America.

This asymmetry in migration success has generated much debate, with the traditional explanation putting the dominance of the North American species down to more advanced evolution and the supremacy of carnivores from the north. This old-fashioned view has been well and truly debunked. The latest dating has shown that many of the South American species died out long before the Panama Isthmus was in place, so they were not outcompeted by North American invaders.

The formation of the Isthmus of Panama in the Pliocene not only changed global ocean circulation and helped to cool the Arctic, it also enabled multiple species migrations (of animals and plants) that reshaped biodiversity in the Americas with long-lasting effects on ecosystems and evolution across both continents. Iconic animals like jaguars, llamas and armadillos owe their current distributions to the creation of this land bridge. The Pliocene world also saw the emergence of cycles of cold glacial and warmer interglacial climates that would become much more pronounced in the great ice ages that dominated the Quaternary Period from 2.58 million years ago.

# CHAPTER 29

# Ocean Currents and Earth's Biggest Beasts

When you stand on a seashore watching the breaking waves, feeling the water running over your feet, you become connected to every part of the global ocean. The water between your toes may have spent time in the Atlantic Gulf Stream, in the darkness of a deep ocean trench, flowing around the Antarctic continent on the edge of the sea ice. Perhaps all of these. We must think of a global ocean because the oceans are connected. Water is cycled around our blue planet via a network of ocean currents; those that flow at the surface are mainly driven by winds while the deep, cold ocean currents are powered by differences in water density.

The oceans are the heart and lungs of our planet. Ocean currents move heat, oxygen and carbon around the world, playing a fundamental role in the climate system and supplying nutrients that fuel the marine food web. Today's network of ocean currents evolved during the Cenozoic Era as the familiar geography of today's world shifted into place. As continents drifted, marine gateways opened while others closed. Tectonic processes rerouted ocean currents in both hemispheres with profound consequences for climate and life.

When the mass extinction at the end of the Cretaceous cleared the Earth of big marine reptiles, a vast ocean of evolutionary possibilities was created. The largest creature in Earth history, the blue whale – 150 tonnes of ocean-going beast – evolved during the Cenozoic. Today there are ninety species of whale, the product of a remarkable evolutionary history that began on land in the Eocene Epoch. The success of these magnificent creatures is intimately linked to the global network of ocean currents. Whales follow their food, and ocean currents control where that food is located.

The ocean plays a key role in regulating Earth's climate because water has a remarkable capacity to absorb large amounts of heat without seeing a large increase in temperature. The upper 3 metres of the ocean can hold more heat than the Earth's atmosphere. Most of Earth's incoming solar energy is absorbed by the upper 50 metres of the ocean, with the greatest warming close to the equator. The tropical latitudes receive the most solar energy and the polar regions the least, so ocean currents work to cancel out this energy imbalance by moving heat from the low to the high latitudes.

This energy is moved polewards by warm surface currents. The Gulf Stream, for example, conveys some 65 million cubic metres of water every second from the Gulf of Mexico to the far North Atlantic. When it arrives in the seas between Greenland and Norway, this water that was already salty due to the moisture lost to evaporation in the warm tropics now becomes cooler as it is chilled by Arctic winds. This cold and salty, and therefore dense, water sinks to great depths and creates what oceanographers call North Atlantic Deep Water (NADW). This flows southwards back towards the equator in the blackness of the abyss, hugging the ocean floor at depths of 2,000 to 4,000 metres. The cold, salty current seeks out the deepest parts of the ocean like a river flowing downhill. The deep ocean basins are filled with water that is close to 0°C.

The formation of NADW is a crucial part of what is known as Earth's *thermohaline circulation*, which is powered by differences in water temperature (*thermo*) and salinity (*haline*). This circulation involves all the oceans and is often called the *global ocean conveyor belt*. In illustrations, the warm surface currents are commonly shown in red, the cold currents of the ocean depths in blue. The huge volumes of water that sink in the North Atlantic, supplemented by waters that sink around Antarctica, create a powerful downward pull that helps to drive this planetary-scale circulation. It is a crucial component of the Earth's climate system, redistributing heat and regulating climate patterns across the globe. It also helps sequester carbon by transporting carbon dioxide-rich water to the deep ocean.

This thermohaline circulation mixes the waters of the global ocean by linking, via a great loop, the major surface and deep-water currents in the Atlantic, Indian, Pacific and Southern oceans. The deep and slow-moving cold water currents eventually return to the ocean surface by a process known as upwelling. This process completes the loop and sustains marine ecosystems. One lap of this global circulation takes a surprisingly long time; it has been estimated that the travel time for deep water formed in the North Atlantic to reach upwelling zones in the Pacific is about 1,600 years.

Where strong surface winds push warm surface water away from an area, it is replaced by cold water from below. Zones of upwelling can be seen from space, where satellites with thermal sensors detect the colder waters emerging at the sea surface. Upwelling zones are among the most biologically productive regions of the ocean, supporting large populations of fish and other marine organisms because deep water is loaded with nutrients and organic matter.

Upwelling is critical because it supplies nitrogen and phosphorus that are essential for the growth of single-celled microscopic plants known as phytoplankton. The ocean depths contain lots of these nutrients from decomposing organic matter. When plants and animals die in the open ocean, some or all of their biomass will eventually sink to the depths and break down. When upwelling brings these nutrients into the zone warmed by sunlight, phytoplankton growth is boosted, with knock-on effects for the entire marine food web. When conditions are favourable, phytoplankton populations can explode to produce a phyto-plankton bloom that may last for several weeks. Satellites can map these blooms (which can extend over hundreds of square kilometres) because the phytoplankton's green pigment changes the colour of the nutrient-rich waters.

A big upwelling enriches the entire food chain. When nutrients fertilise the surface waters it means the marine environment can support more zooplankton like krill, which in turn provides a food source for juvenile fish and jellyfish, right up to the biggest creatures on Earth. Large shoals of sardines, for example, attract predators like bluefin tuna and sharks. Strong upwelling events can produce spectacular concentrations of marine life; in summer 2024, seventy blue whale sightings were made off the southern coast of Australia within an area of just 10 nautical miles.

Upwelling is a good example of an interaction between Earth's atmosphere and ocean that has profound biological consequences, creating biodiversity hotspots. The ecological richness of the Galápagos archipelago – a place that inspired many of Darwin's ideas around evolution and the creation of new species – is a direct function of the seasonal upwelling of unusually cold nutrient-charged waters to the west of the islands. This upwelling is driven by strong winds in the eastern equatorial Pacific that mix the surface waters and draw cold, deep waters to the surface. These waters are extremely rich in marine algae, which provides food for many animals. Marine iguanas are endemic to the Galápagos archipelago, where they forage for food in the ocean (the only lizards that do so). They graze on the red and green algae that grows on the rocky coast and in deeper water. This food source is critical for their success.

The Southern Ocean that encircles Antarctica is the largest feeding ground for marine mammals. Each summer, oxygen- and nutrient-rich currents stimulate intense bursts of phytoplankton production powered by nearly twenty-four hours of sunlight each day. The blooms attract vast numbers of zooplankton, and upwelling creates feeding hotspots, attracting dozens of species of whales, dolphins and other creatures to feast in the Antarctic waters as part of one of the densest concentrations of life on Earth. The seasonal migration patterns of many species are fine-tuned to align with the timing and location of shifts in marine productivity driven by the major ocean currents.

In Chapter 18 we explored one of the most important developments in the history of life – when four-limbed animals left the ocean and walked on dry land in the Devonian Period. The first tetrapods were probably amphibious animals, spending time both on land and in water before finally leaving an aquatic life behind. In the Cenozoic Era, some four-legged animals went in the other direction, swapping life on land for life in the oceans. All whales, dolphins and porpoises (collectively known as cetaceans) are descended from four-legged mammals that once lived on land.

The evolution of whales is a fascinating journey that spans over 50 million years, transforming these mammals from land-dwelling creatures to the fully aquatic giants of the oceans. It is an evolutionary process that highlights a raft of anatomical and behavioural adaptations to marine life. The earliest whales lived in the Eocene and were amphibious. Eocene fossils from India and Pakistan trace the origin of cetaceans to a small, four-legged deer-like animal that lived in southern Asia over 50 million years ago. This was a land-dwelling animal that also waded in shallow water, foraging for small fish and invertebrates, and may have swum out to deeper water to avoid predators. During the Oligocene Epoch, the front limbs evolved into flippers for steering, while the hind limbs regressed into vestigial structures and eventually disappeared. By the Miocene, whales had developed long, streamlined bodies and a powerful horizontal tail fluke for propulsion in water. Their nostrils had migrated to the top of the skull to form a blowhole, allowing them to breathe easily while swimming. They also developed specialised ear structures for underwater hearing. Some whales, like orcas and sperm whales, developed teeth, while species such as blue whales and humpbacks developed baleen plates for filter-feeding krill and small fish.

This evolutionary journey highlights how dramatic environmental changes and ecological opportunities can drive significant adaptations, leading to the emergence of entirely new ways of life. Cetaceans are mammals but they cannot survive out of water. They possess a variety of features that inform us about their distant land-based heritage. They have lungs and breathe air at the water surface. All whales, dolphins and porpoises lack hind limbs, but these limbs begin to form in the early weeks of embryo development before disappearing as the embryo grows. Genetic studies have revealed links between cetaceans and a group of animals known as the *artiodactyls*, or even-toed ungulates. Today this includes camels, deer, pigs, sheep, giraffe and hippos. The closest living relative of the cetaceans is the hippo. They share 'pod' as a collective noun.

Where zones of upwelling produce hotspots of marine life, they attract large predators, like orcas and sharks. The Miocene ocean saw the emergence of a fearsome marine predator known as megalodon (*Otodus megalodon*) – you may have seen the 2018 Hollywood blockbuster *The Meg*, based on the 1997 sci-fi novel *Meg: A Novel of Deep Terror*. Megalodon means 'big tooth' – its teeth are nearly three times larger than those of a modern great white shark.

Megalodon was not a Hollywood fiction – it was probably the largest fish that ever lived. There has been much speculation about the size of megalodon because sharks are poorly represented in the fossil record – their skeletons are made of cartilage, which does not survive burial as well as bone. But we do have millions of well-preserved shark teeth. Megalodon teeth have been found in marine deposits on every continent apart from Antarctica. It is possible to estimate the size of the shark body from the size of its teeth by making comparisons with the anatomy of living sharks.

Megalodon was a global apex predator; perhaps the largest marine predator of all time. It was huge – a full-grown adult could be 15 to 20 metres long, which is about three times longer than the largest known great white shark. It dominated the Miocene and Pliocene oceans for about 20 million years, hunting whales, seals, large fish and even other sharks. Imagine this beast tracking warm currents into upwelling zones. There are fossil whale bones with megalodon tooth marks, including bite patterns close to fins, suggesting strategic attacks. The megalodon died out at the end of the Pliocene Epoch when Earth entered a phase of global cooling and ice-sheet growth lowered sea levels. We cannot be certain about the cause of the megalodon extinction but a shrinking habitat (they preferred warm tropical waters) and a decline in their food supply were probably important. Megalodon co-evolved with whales but, while the latter developed a thick layer of blubber for insulation, megalodon could not cope with the icy waters of the high latitudes.

One of the reasons for megalodon's extinction may have been the abrupt global cooling when sea-surface temperatures fell 2.8 million years ago. We do not find any megalodon fossils after this time. Megalodon may have lost a major part of its diet by not being able to follow its prey to upwelling zones in colder waters. A number of large marine animals, including several species of sea turtles, went extinct at around the same time. This signalled the end of the bonanza for the ocean's biggest ever shark.

# CHAPTER 30

# Out of Africa: The Human Story

About 3.7 million years ago, in the Pliocene Epoch, a volcano erupted explosively in the East African Rift Valley, covering the landscape with a layer of fine grey ash. At Laetoli in northeast Tanzania, two adults and a child walked across the ash, leaving a trail of footprints. The child followed the adults, placing tiny feet beside their prints. After gentle rain the footprints hardened before another fall of ash sealed them in time. The beautifully preserved Laetoli footprints occupy a special place in the human story because they provide the oldest unequivocal evidence of upright walking on two legs.

It all started between 7 and 5 million years ago, as Earth continued the trend of gradual Cenozoic Era cooling. Important changes took place in African ecosystems in response to the regional climate becoming cooler and drier. Rainforests contracted, allowing open grasslands and savannah environments to expand. As forest cover diminished, tree-based resources and shelter became scarcer, encouraging early hominins (the group that includes modern humans, extinct human species and all our immediate ancestors and relatives) to spend more time on the ground and in more open landscapes. Many primates, including chimpanzees and bonobos, remained in the trees or adopted semi-terrestrial life-styles, while hominins evolved the ability to walk upright on two legs and use tools. Apes do this too, of course, but we took tool making and planning to another level. These traits allowed hominins to navigate open landscapes and exploit new ecological opportunities.

This new way of life favoured hominins who could move efficiently on the ground, avoid predators and locate dispersed sources of food. Walking on two legs allowed mothers to carry offspring while foraging for food, a key adaptation for species with young who take more than a year to become independent. By evolving this adaptability and flexibility, the human lineage was able to thrive in a range of environments. This set the stage for some hominin groups to leave Africa and for the later evolution of modern humans and their dispersal around the world.

The late Pliocene Laetoli footprints were discovered in 1976 and fully excavated two years later by a team led by the renowned British palaeoanthropologist Mary Leakey (1913–1996). Palaeoanthropology is the scientific study of the origins, evolution and development of early humans and their ancestors, which draws from many fields, including archaeology, biology, Earth science and genetics. The trackway Leakey and her team discovered is 70 metres long and there is widespread agreement that the hominin *Australopithecus afarensis* made the footprints. This name means 'southern ape from Afar' because many fossils of this hominin have been found in the Afar region of northeast Ethiopia, which is part of the Great Rift Valley. The Rift Valley of East Africa is a vast corridor of rocky escarpments, deep gorges, volcanoes, lake basins and semi-arid plains. This dynamic landscape mosaic is the only place on Earth where humans have always been present since they first emerged.

Excavations in East Africa have unearthed fossils from more than 300 individuals of *Australopithecus afarensis*. The latest dating shows that this species survived for the best part of a million years in Ethiopia, Kenya and Tanzania between about 3.85 and 2.95 million years ago. *Australopithecus afarensis* overlaps in time with the Laetoli footprints. It is pivotal in the human story because some scientists think it was ancestral to several hominin species, including later *Australopithecus* species and the genus *Homo*, while others consider it to be a side branch. We cannot claim direct ancestry, but every human alive today is related to the hominins who walked across the bed of ash at Laetoli 3.7 million years ago.

The most famous *Australopithecus afarensis* individual was discovered in 1974 in the arid badlands of northern Ethiopia, when American palaeoanthropologist Donald Johanson (b. 1943) spotted a bone poking out of a dry gully. It was one of forty-seven bones his team recovered belonging to a single individual who walked the Afar landscape about 3.2 million years ago. She caused a sensation. At the time she was the oldest and most complete human relative yet discovered. She was named Lucy after the Beatles' song 'Lucy in the Sky with Diamonds', which was played loudly in the expedition camp as they celebrated her discovery.

Lucy transformed our understanding of human evolution because she walked on two legs and had a small skull. While she was only about 110 centimetres (3 feet 7 inches) tall, her teeth showed that she was an adult, so it could be established for the first time that walking on two legs happened long before the evolution of large brains. Her anatomy was built for an upright posture and bipedal locomotion; she had lost the key anatomical features needed for life in the trees, with pelvis and leg bones identical in function to those of *Homo sapiens*. More than five decades after their discovery, the forty-seven bones of Lucy's skeleton remain a key reference point for the study of human evolution.

Fossil skulls and teeth are especially important finds because they contain diagnostic features that allow species of early humans to be differentiated. The study of tooth form and wear provides clues about what our ancestors ate. The earliest hominins lived in forest habitats and were adept tree-dwellers and upright walkers. Hominin species in the genus *Australopithecus* were able to live in a range of habitats across Africa. Their teeth indicate consumption of a broader range of food types than their ape ancestors, whose diet was predominantly fruit. Jaw and dental features suggest *australopiths* were able to eat hard-to-process foods such as nuts, seeds and roots. Later hominins in the genus *Homo* had large brains and smaller teeth and jaws. They developed the ability to manufacture sophisticated tools for hunting animals and for processing a wide variety of foodstuffs.

*Homo habilis* (handy human) was discovered by Louis and Mary Leakey and their son Jonathan in the Olduvai Gorge in 1960. It was initially thought to be the first hominin to use stone tools (simple chipped pebbles). *Homo habilis* lived in East and perhaps southern Africa between about 2.0 to 1.6 million years ago, overlapping with *Homo erectus* for some half a million years. *Homo erectus* (upright human) emerged about 2 million years ago and was the first of the hominin species known to have left Africa. It is also the oldest known hominin to have a body essentially like ours, with long legs and shorter arms relative to its torso. It was built for walking and running on two legs. The fossil record of *Homo erectus* spans well over 1.5 million years, with the last evidence of its presence in Southeast Asia some 110,000 years ago. The most complete and best-preserved skulls attributed to *Homo erectus* have been found in Europe at the Dmanisi archaeological site in Georgia, which dates back some 1.8 million years. *Homo erectus* was the first human species to manufacture the more complex tools we call handaxes, from about 1.7 million years ago. For much of this time, stone tools of various shapes and sizes were used to process plant foods and for butchering carcasses – a period that archaeologists call the Palaeolithic (Old Stone Age).

*Homo sapiens* means 'wise human', and every human alive today belongs to this species. The name was introduced in 1758 by the Swedish taxonomist Carl Linnaeus as part of his grand system for classifying living organisms. It was chosen to highlight the intellectual and cultural traits that distinguish humans from other animals. There is still much uncertainty about where and when *Homo sapiens* emerged because the fossil record is patchy and some sites are not well dated. The earliest known fossil remains of *Homo sapiens* were first discovered in the 1960s in a limestone cave at the Jebel Irhoud site in western Morocco, which came as something of a surprise given that research in Africa has traditionally been focused on sites in the east and south of the continent. The remains of twenty-two hominins have now been recovered from the cave, and with recent improved dating methods are now classified as early *Homo sapiens*, dating from around 300,000 years ago.

Groups of *Homo sapiens* left Africa in several waves; the earliest may have been over 150,000 years ago. These migrations did not result in viable long-term populations being established until about 50,000 years ago, when our species began spreading across the globe, replacing or interbreeding with other hominin species such as Neanderthals (*Homo neanderthalensis*). Key to this global dispersal was an ability to adapt to changing environments.

Between 50,000 and 40,000 years ago, during the last glacial period, three human species lived in Europe, western Asia and parts of Siberia. Denisova Cave in the Altai Mountains of southern Siberia has provided extraordinary insights into the human story at this time. In 2008, a finger bone from a juvenile hominin was discovered in a sedimentary layer in the cave dated to between 76,000 and 52,000 years ago. DNA extracted from this tiny bone led to the identification of a completely new group of hominins now known as the Denisovans. A tooth found in the cave contained a remarkably similar genome.

Denisova Cave is hugely important because we have evidence of three distinct human groups using the cave: Denisovans and Neanderthals episodically before about 45,000 years ago and modern humans after that date. Excavations at the cave have revealed a variety of stone tool types made by Neanderthals, Denisovans and modern humans. Artefacts discovered in the cave include a fragment of ostrich egg, a polished greenstone bracelet, a cave lion figurine carved from woolly mammoth ivory and a bird bone needle dated to about 50,000 years ago – the oldest needle ever found.

Neanderthal fossils have been found in hundreds of locations across Eurasia, and we have excellent data on their anatomy, yet fossil remains of the Denisovans are extraordinarily rare. At the time of writing, the global collection includes one finger bone and some teeth from Denisova Cave itself, a jaw bone from the edge of the Tibetan Plateau, a tooth from a cave in northeastern Laos, and a jaw bone from Taiwan, but it is suspected that several enigmatic Chinese fossils may turn out to be Denisovans. The current scarcity of Denisovan fossils is in stark contrast to the spectacularly rich datasets from ancient DNA, which can be well preserved in fossil bones and sediments buried in cool, damp cave environments, but it does not last long after burial in the warm climates of Africa. It is a method that is yielding remarkable detail on the human story.

The Neanderthals were stocky, muscular and robust. They navigated the great climate shifts of the Quaternary ice age for over 350,000 years, first appearing in the fossil record some 400,000 years ago. In many respects they were hugely successful, yet they have long been mocked as a brutish and stupid genetic dead end, the archetypal club-wielding cave man. Recent years have seen a major reappraisal of their cognitive abilities and social complexity. Neanderthals had art, they buried their dead, they showed remarkable dexterity in crafting elaborate stone tools, they cared for their sick, they foraged for shellfish, seeds, nuts and a wide range of plants. They hunted the mighty woolly mammoth and mastered the use of fire.

Neanderthals lived in Europe and Asia for hundreds of thousands of years before *Homo sapiens* arrived on the scene. DNA evidence shows that Neanderthals bred with both *Homo sapiens* and Denisovans. The latest genetic research shows that modern humans and Neanderthals last interbred in a brief ice age window between about 50,000 and 45,000 years ago. Neanderthal populations had a long time to adapt to life outside Africa so when they bred with modern humans, they passed on some of their acquired advantages, such as an immune system and skin pigmentation that performed better in cold climates. The ancestors of all the people living today who carry some Neanderthal DNA (averaging about 2 per cent outside of Africa) obtained that DNA during this ice age window. In fact, there is more of their DNA on Earth today than at any time in the ice age; only those whose ancestry comes solely from sub-Saharan Africa have little or no Neanderthal DNA. So, while the Neanderthals vanish from the fossil record about 40,000 years ago, they do not disappear completely. *Homo sapiens* took their DNA with them as they migrated to all parts of the world.

Before about 1960, most palaeoanthropologists believed that only a single hominin species existed at any given time in the past and the evolutionary story – like the famous 'ascent of man' T-shirt – was a more-or-less linear progression of one species of human ancestor replacing or evolving into another. As dating control has improved, it has become clear that several species of hominins co-existed for most of the last 5 million years. The study of ancient DNA from the last 60,000 years in Europe and Asia is revealing that gene flows between co-existing archaic human groups were the norm. This was surely the case in Africa too, at least for closely related species.

There are still large gaps in the hominin fossil record and the precise evolutionary link between modern humans and our early ancestors is uncertain. But it has been established from fossil discoveries inside and outside Africa that the human family tree is much more complex than previously thought, with multiple stems and branches. It is not possible to mention all of the hominin species here, but we should think of ourselves as a small twig at the end of one of these branches since, in terms of time spent on this planet, *Homo sapiens* account for barely 5 per cent of the hominin story.

# CHAPTER 31

# The Quaternary Ice Age

Take a walk through Central Park in New York and at every turn you will see the work of ancient ice. The hard Manhattan schist – the bedrock beneath the city's dizzying skyscrapers – has been moulded, grooved, polished and scratched by glacial action. By picnic spots and pathways, granite boulders sit stranded where an ice sheet in retreat set them down.

Between 25,000 and 18,000 years ago, Canada and the northern fringes of what is now the USA, including all of the Great Lakes, were buried under the Laurentide ice sheet; New York sits just inside its southeastern margin. At its centre near Hudson Bay, this mountain of glacial ice was over 3 kilometres thick; it was joined to the Cordilleran ice sheet to the west of the Rockies and to the Greenland ice sheet in the far north. At the peak of the last glacial period, after ten thousand winters of snow, there was more glacial ice in North America than today's Antarctica.

The ice age legacy is evident across this vast region and beyond. As glacial sands and gravels piled up at the ice sheet margins, terminal moraines formed with ridges over 100 metres high. The Harbor Hill Moraine on Long Island is a major landscape feature running parallel to today's northern coast. Without this glacial barrier the boroughs of Queens and Brooklyn would lie beneath the waters of the Atlantic Ocean.

The advance and retreat of ice sheets in Europe and North America is the defining feature of the Quaternary Period, the last period in our Earth history beginning 2.58 million years ago and extending to the present day. It is the geological period that saw the emergence of *Homo sapiens* (Chapter 30) and in which we still live. The Quaternary is punctuated by great swings in climate as conditions alternated between long cold glacials and much shorter and warmer interglacials. It is formally subdivided into the Pleistocene (2.58 million to 11,700 years ago) and Holocene (11,700 years ago to the present) epochs – the Holocene is the present interglacial.

As well as providing the backdrop to the evolution of our species, these shifts in Quaternary climate were accompanied by profound changes to landscapes and ecosystems. In glacial Britain, the steppe-tundra environment beyond the ice was inhabited by cold-adapted species such as reindeer, woolly mammoths and Arctic foxes. Ice sheets reorganised drainage patterns – the largest glaciation in Britain diverted the River Thames southwards to its present course. Very differently, the last interglacial was warmer than today and saw hippos (*Hippopotamus amphibius*) swimming in the Thames, with much of Europe densely forested.

When the Quaternary ice sheets advanced in North America and Europe, global sea levels fell dramatically as water was taken from the oceans and locked up as glacial ice on the continents. By the time of the last glacial maximum, around 22,000 years ago, global sea level had plunged some 120 metres lower than today so that great expanses of continental shelf became dry land. These environmental changes created an ice age supercontinent as the Americas became connected to Eurasia and Africa.

This lowered sea level transformed global geography, connecting formerly separated landmasses via land bridges, allowing people, plants and animals to migrate to new places. Eastern Siberia was joined to Alaska via the Bering land bridge, Britain was connected to Ireland and mainland Europe, the Persian Gulf was dry land, Japan was linked with mainland Asia, and an ice age continent of some 10.6 million square kilometres, known as Sahul, incorporated the landmasses of Australia, Tasmania and New Guinea. During the course of the last glacial period, between about 80,000 and 12,000 years ago, these connections – perhaps with some island-hopping to Australia – allowed humans to reach every continent apart from Antarctica. The sea-level fall also joined up Quaternary Earth's largest biome – the mammoth steppe. This biome supported more animal and plant life than today's African savannah.

The woolly mammoth (*Mammuthus primigenius*) is the iconic beast of the Quaternary ice age. It had curved ivory tusks that could grow over 2 metres in length. This animal was remarkably well adapted to the cold glacial climate. It had a thick layer of insulating fat beneath its tough leathery skin, and all parts of its body, including head, ears and trunk, were covered by fur that ranged in length from a few centimetres to almost a metre. Even the woolly mammoth's feet and toes were shrouded in fur. On top of this fat and fur its giant frame was covered by an overcoat of long, thick, wiry guard hairs. Unlike modern elephants, the woolly mammoth had small ears, another evolutionary adaption to limit heat loss.

A fully grown adult woolly mammoth was about the same size as an African bull elephant, with a shoulder height of about 3 metres and a weight of between 4.5 and 6 tonnes. It would have consumed upwards of 350 kilograms of plant matter every day. This remarkable beast thrived in a habitat that no longer exists. Many parts of Europe, North America and Siberia that were not covered by glacial ice were carpeted by a highly productive herb-dominated grassland vegetation that supported a mix of large grazing animals. This cold, dry biome is known as the mammoth steppe. During the Pleistocene glacials it wrapped around the northern hemisphere in a more-or-less continuous belt, from the Atlantic coast of Europe to eastern Russia and across the Bering Strait into Alaska. When there was a corridor between the Cordilleran and Laurentide ice sheets, it extended right across

North America to the Atlantic coastal plain. This mammoth super-highway was populated by millions of large herbivores, including woolly rhinos, musk oxen, bison and horses.

The mammoth steppe was a highly productive biome dominated by flowering herbs, grasses and dwarf willow. This was the largest wildflower garden of all time; it supported a very wide variety of plants including sedge, goosefoot, buttercup, campion and sagebrush. In 1901, near the Berezovka river in the far east of Russia, a well-preserved woolly mammoth carcass was discovered with buttercups and grasses lodged between its teeth. It had been buried in the deep freeze of the Siberian permafrost for more than 40,000 years. Apart from a few patches, this biome is now largely extinct. Much of Siberia, for example, is warmer than the mammoth steppe in the glacial periods and the soils tend to be boggy and less fertile.

There are conflicting views on the reasons for the decline of this biome. Some researchers have argued that as the climate became wetter at the end of a glacial period, the expansion of woody shrubs and peatlands led to the loss of this habitat, and the big grazers departed. Others have argued that the extinction of these big grazers led to an increase in woody shrubs and the degradation of the mammoth steppe biome. Whatever the reason, woolly mammoths went extinct on the Siberian and North America mainland at the end of the last glacial period as a combination of climate changes, habitat loss and human hunting sent them over the edge. But remarkably, on Wrangel Island in the Arctic Ocean, off the far northeastern coast of Siberia, a population of woolly mammoths survived into the second half of the Holocene. Radiocarbon dating shows that they finally died out about 4,000 years ago, outliving the mainland populations by some 7,000 years. Woolly mammoths were living on Wrangel Island when the bluestones were being erected at Stonehenge on Salisbury Plain.

Ice age science emerged in the nineteenth century as old ideas about the biblical flood were cast aside and geologists began to recognise fossils of cold-adapted species and evidence for the work of glaciers in places that were great distances away from the nearest modern glacier. A late summer field trip in Scotland in the middle of the century became a landmark in our understanding of Earth's glacial history. On 23 September 1840 William Buckland and Louis Agassiz left Glasgow by stagecoach on a tour of the Scottish Highlands in search of evidence for the work of glaciers. Agassiz was an expert on fossil fish, but after spending the previous few years studying the work of glaciers in the Swiss Alps, he had developed a grand theory of an ice age epoch with an ice sheet covering much of the northern hemisphere. This tour of the Highlands was a great success – the two men saw evidence for glaciation almost everywhere they looked. Their excursion was a milestone in the history of geology because it led to the first report of the work of ancient glaciers in a country where glaciers were absent today.

These discoveries sparked new debates in Europe and North America about climate change and the extent to which the actions of glaciers had been important in shaping the landscape. By the 1870s most British geologists accepted the glacial theory but there was much disagreement about the extent of these glaciers. Leading naturalists like Lyell and Darwin were happy to recognise the action of valley glaciers in the mountains of Scotland and Wales but were not convinced that an ice sheet had covered much of Britain and Ireland. By the end of the nineteenth century, however, after careful mapping of glacial deposits and landforms, most geologists accepted that an ice sheet had formed, and the debate turned to how many glacial episodes had taken place. How many times did large ice sheets develop and waste away during the Quaternary Period?

Field research in the Alps in the early twentieth century found evidence of four glacials and four interglacials during the Quaternary. Two of the leading geomorphologists of this era mapped river terraces and glacial moraines in the Alpine foothills and found evidence for four glacial phases. Albrecht Penck (1858–1945) and Eduard Brückner (1862–1927) published their findings in 1909 in a major three-volume work, *Die Alpen im Eiszeitalter* (*The Alps in the Ice Age*). Their model of four glacials dominated ice age science for much of the twentieth century.

The geological record in the Alps and in other regions that were scoured by large glaciers is very patchy because the erosive action of glacial ice sweeps away the deposits from the earlier glacials and interglacials. The geological record is full of holes so key chapters in the ice age story are missing. We have to look elsewhere for a complete record of the environmental shifts of the Quaternary Period. The fine sediments and tiny fossils that accumulate on the floor of the deep oceans preserve a remarkably detailed unbroken record of change that spans the entire Quaternary Period. Sediment cores from this environment provide a continuous record of global climate changes through the Quaternary ice ages because the deposits accumulate slowly, millimetre by millimetre, in a low-energy environment that is not compromised by erosion.

Modern analysis of the marine sediment record has revealed some fifty glacial and interglacial cycles within the Quaternary Period. This work showed that Penck and Brückner's Alpine scheme of four glacial periods and four interglacials was far too limited. The study of the marine sediment record has revolutionised understanding of Earth history, demonstrating that the environmental changes of the Quaternary ice ages were much more frequent and rapid than anyone could have previously imagined. It also demonstrated that the continental-scale reorganisation of ecosystems from glacial to interglacial and back again must have taken place many more times than previously thought. The repeated expansion and contraction of the mammoth steppe biome, for example, suggests a megafauna that was adapted to environmental change. The big difference at the end of the last glacial period, when many species went extinct, was rapid ecosystem change alongside the widespread presence of well-drilled human hunters.

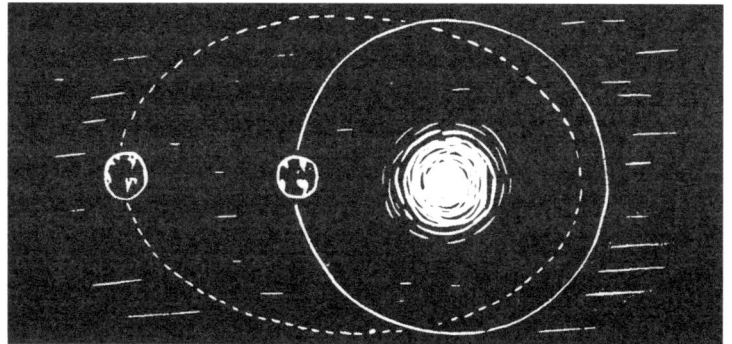

# CHAPTER 32

# Seasons and Celestial Cycles

When dawn breaks on the winter solstice at the prehistoric Newgrange tomb in Ireland, the narrow entrance passage directs the Sun's rays to a central chamber where elaborately carved slabs of rock are bathed in sunlight. If the midwinter sky is clear, the inner chamber is illuminated for almost seventeen minutes. Newgrange is the world's oldest known astronomically oriented structure – it was engineered by Neolithic farmers over 5,200 years ago to mark the beginning of a new year and the gradual return to the longer, brighter, warmer days of spring and summer.

The Sun is the primary source of energy driving Earth's climate, but its strength has not been constant. Nor has the length of the days and seasons. The Sun's brightness has slowly increased over the last 4.5 billion years as more hydrogen is converted to helium in its core. Today our star produces about one-third more heat than it did when Earth was formed. Astronomical factors such as the tilt of the Earth and the shape of its orbit around the Sun influence the character of the seasons and how much of this solar radiation is received across the Earth's surface. One of the great achievements of modern Earth science has been the detection of celestial rhythms in the geological and archaeological records. Since the Earth's rotation wobbles over a 23,000-year timescale, the Sun's rays now enter the passage at Newgrange about four minutes late. Five thousand years ago Neolithic people would have witnessed first light on the winter solstice entering exactly at sunrise.

Seasons are experienced very differently across the Earth's surface. If you live close to the equator you will see only limited seasonal change – a rainy and dry season – but the climate stays warm, and the length of the day does not change much throughout the year. The Arctic and Antarctic on the other hand experience extreme seasonal change in day length and the climate is cold all year round. The Arctic is the land of the midnight sun – in summer it does not get dark at the North Pole while in the Antarctic winter the Sun does not appear above the horizon for several months. The zones between the tropics and the polar regions in each hemisphere are the mid-latitudes. This is where the distinctive character of all four seasons is most clearly expressed and where the living and physical worlds respond accordingly.

We have seasons because the Earth's axis of rotation is tilted with respect to its orbital plane. In other words, the imaginary line passing through both poles is not perfectly lined up with the Sun. As the year progresses, parts of our planet receive quite different amounts of solar radiation. The northern hemisphere is tilted towards the Sun in summer. This means the Sun is higher in the sky, so days are longer and warmer (more hours of sunlight). The northern hemisphere is tilted away from the Sun in winter. This means the Sun is lower in the sky, so days are shorter and cooler (fewer hours of sunlight). These timings are reversed in the southern hemisphere, where the winter solstice is in June and the summer solstice in December. It is highly likely that the Earth's tilt is a hangover from the catastrophic collision with a planet-sized object about 4.4 billion years ago that led to the formation of the Moon, so seasons have been around for most of Earth history.

Earth completes a full rotation every twenty-four hours, giving us day and night. This daily rhythm signals to animals and plants when it is time to rest and when it is time to be active. We take it for granted that there are twenty-four hours in a day and 365 days in a year. But early in Earth history days were shorter because the Earth's rotation was faster. About 3.5 billion years ago, in the Archean Eon, a single rotation took about nineteen hours. When the Cambrian explosion took place some 3 billion years later, day length had increased to about twenty-one hours. There would have been 417 days in the early Cambrian year. Earth's rotation has slowed down over time because of friction effects caused by the tidal movement of the oceans over the sea floor and because the Moon has moved further away and captured some of Earth's rotational energy. Every hundred years these processes add about 2.3 milliseconds to the length of a day. In about 200 million years from now a single day will last for twenty-five hours. The shorter days of the Archean would have influenced many aspects of early Earth, including tidal cycles, the warming of the atmosphere and possibly the evolution of life.

The increase in day length has not been constant over geological time. Between about 2 and 1 billion years ago, in the Proterozoic Eon, day length seems to have been stuck at about nineteen hours, coinciding with what some Earth scientists have called the 'boring billion', when the oxygen content of the atmosphere stalled after an initial increase. Atmospheric oxygen levels may have flatlined in this period because the amount of oxygen generated in the oceans by photosynthetic bacteria was hampered by shorter days with shorter bursts of sunlight. This idea raises the intriguing possibility that the evolution of the Earth's rotation and day length may have played a role in the early composition of the atmosphere and in the evolution of complex life. Only when oxygen production had ramped up was it possible for more complex life to evolve.

The living world responds to seasonal and annual cycles in myriad ways. Have you ever counted the rings on a tree stump? Outside the tropics, trees form a ring every year because they stop growing in the winter. Trees produce a thick ring when the growing season is favourably warm and wet, but when the weather is too cold or too dry the rings tend to be thinner. Changes in the density and chemistry of the wood in each ring can also tell us about the climate in the year the ring was formed. By linking the tree-ring record from living trees to data on temperature and rainfall collected from the nearest weather station, we can relate how the tree has grown to the year-to-year fluctuations in climate. From this relationship, we can work backwards with tree-ring records from old wood to build up a picture of climate in the past. We can cross-match and overlap records from living and dead trees, wooden beams in old buildings, even the masts from Viking ships and blackened oaks pulled from ancient bogs. The study of tree-ring data from sites around the world is one of the methods that allows us to put the monitored climate change record from the last century into long-term context.

Seasonal changes in Earth processes can also be recorded in the sediments deposited in glacial lakes. Early in the twentieth century they were used to build the first precise timescale from geological sections. Gerard De Geer (1858–1943) was a Swedish geologist from an aristocratic family – his father was Sweden's first prime minister. He is famous for recognising that the thick stacks of fine sediment that were deposited in a huge Baltic lake close to the retreating margin of the last Scandinavian ice sheet were laid down in distinctive annual layers. He called these *varves*, after the Swedish word for 'layer'.

During every spring and summer meltwater season, streams transported sediments to this lake and they were deposited on the lake bed. The heavy sands were deposited first to form a distinctive gritty summer layer while the fine silts and clays remained suspended in the deep water column before finally settling out during the autumn and winter. This produced an annual couplet – a distinctive two-part layer with sharp upper and lower boundaries that represents approximately one year of sedimentation. De Geer saw great stacks of varves all across Sweden in railway cuttings and in valleys where rivers had cut through them. He realised that each varve represented one full year (summer plus winter) of deposition, and he began to count them. His year-by-year record extended all the way back to the end of the ice age over some 12,000 years.

While De Geer was counting thousands of annual varves in Sweden, a brilliant Serbian mathematician was contemplating celestial cycles for a grand theory of how changes in the Earth's position relative to the Sun could change the climate and trigger an ice age. Milutin Milankovitch (1879–1958) examined the shape of the Earth's orbit around the Sun, which because of the gravitational pull of Jupiter and Saturn varies from circular to an ellipse over a period of almost 100,000 years (*eccentricity cycle*). He also studied the changing tilt of the Earth's axis with respect to Earth's orbital plane (*obliquity cycle*) and how the direction of the Earth's axis of rotation wobbles over time (*precession cycle*). These became known as the Milankovitch cycles. With obsessive dedication over three decades, he calculated how these cycles could influence the amount of solar radiation (known as *insolation*) received at the Earth's surface.

Over the last million years or so Earth's angle of tilt has varied between 22.1 and 24.5 degrees with respect to its orbital plane around the Sun. When the tilt has shifted from one extreme to another and back again (this takes 41,000 years), an obliquity cycle has been completed. The precession cycle operates over a 23,000-year cycle: as Earth spins, its axis of rotation wobbles in a circle like a spinning top slowing down. The more the Earth is tilted the more extreme our seasons become, because insolation is greater in summer and less in the winter. When the angle of tilt is at its minimum, there is less contrast between summer and winter and summers become cooler. The eccentricity cycle meanwhile changes the length of the seasons. At present, northern hemisphere summers are about four days longer than winters, but when the orbit is circular the seasons are the same length.

Milankovitch argued that the amount of solar radiation received at a latitude of 65°N (which is just south of the Arctic Circle) was a key influence on climate. Moreover, he showed that solar energy input at this latitude could vary by as much as 25 per cent. His calculations showed how the three orbital cycles could work in tandem to reduce the amount of solar energy received in summer.

This is critical for sustaining glaciers in the mid-latitudes because summer cooling allows snow and ice to survive the summer melt season so that it can accumulate year on year to eventually build a huge ice sheet. Feedbacks are involved because as more snow and ice survives, the white surfaces reflect more solar energy back into space to promote further cooling and this cooling encourages the build-up of more snow and ice.

Milankovitch's ideas could not be fully evaluated in his lifetime because there were no suitable long records of environmental change with good dating control. In 1976, almost two decades after Milankovitch's death, a study of the climate record preserved in deep-sea sediment cores revealed that the Milankovitch cycles corresponded with major global climate changes during the Quaternary Period. It was demonstrated for the first time that the growth and retreat of the great ice sheets was paced by orbital cycles since they controlled the input of solar radiation and its seasonal distribution. The realisation that the gravitational pull of the giant planets could control the timing of glacial periods on Earth was a profoundly important step in our understanding of Earth history and long-term climate change.

The Milankovitch cycles have operated throughout Earth history during periods of warm and cold climate exerting a fundamental control on climate variability over tens to hundreds of thousands of years. We know that Earth's seasons have danced to the Milankovitch rhythms since the earliest times because the orbital cycles have been detected in banded iron formations from the Archean Eon that are over 3 billion years old. They have also been linked to the advance and retreat of global ice sheets during Snowball Earth events, which suggests the Earth was not completely ice-bound at these times.

Twentieth-century geology was famously reluctant to accept bold new ideas but, like his contemporary Alfred Wegener, Milutin Milankovitch was years ahead of his time. His exhortation to look to the night sky to account for some of the fundamental processes in Earth history was nothing short of a revolution. It is still a timely reminder of our deep connection to the natural rhythms of the universe.

# CHAPTER 33

# Earth History in Ice Cores

Trapped deep within Earth's ice sheets there are bubbles of air once breathed by Neanderthals and woolly rhinos. Because snow very rarely melts in the middle of an ice sheet, it builds up in an orderly fashion year after year, layer by layer, like a sedimentary rock. Each layer begins as freshly fallen snow that is slowly squeezed and compacted into hard glacial ice. The ice stores information about the climate and the composition of the atmosphere when the snow that created it fell to the ground. In this way, ice sheets can provide an unbroken stack of Earth history that can stretch back in time for a million years or more.

To read the history in the ice, we need to drill into the ice sheets and recover long cores of ice. Drilling into an ice sheet poses formidable logistical challenges because the best drilling sites – with the longest records – are located in some of the most inhospitable places on Earth. Ice sheets are dome shaped, with the thickest ice in the highest parts of the deep interior. The Summit Camp on the Greenland ice sheet is 3,216 metres above sea level, where winter temperatures can plunge to 60°C below zero.

Danish scientist Willi Dansgaard (1922–2011) discovered that ice cores drilled from ice sheets contained detailed information about past climates. His pioneering work opened a new window into recent Earth history. Dansgaard studied physics and biology at the University of Copenhagen and worked for the Danish Meteorological Institute. In the summer of 1952 he made a key discovery that shaped the rest of his career and provided a new tool for the study of Earth's past climate. Dansgaard established that it was possible to work out the temperature of clouds by measuring isotopes of oxygen in the rainwater and snow they produced. Dansgaard tested this relationship using temperature data and samples of rainwater and snow from around the world – he even collected rainwater with funnels and beer bottles in his back garden in Copenhagen.

A molecule of water ($H_2O$) contains two hydrogen atoms bonded to a single oxygen atom. Oxygen has three stable isotopes and two of them ($^{16}O$ and $^{18}O$) are commonly used in the study of past climates. Dansgaard found that warmer clouds produced rain with a slightly higher amount of the heavier isotope, oxygen-18, than colder clouds did. The isotopic differences are tiny, but they can be measured with specialist laboratory equipment known as a mass spectrometer.

Dansgaard also established that it was possible to reconstruct past temperatures by analysing the oxygen isotopes in water frozen as ancient glacial ice. He showed that ice cores could provide an annual record of the changing temperature of the atmosphere above the polar ice sheets that reached back in time for many thousands of years.

In 1966 the US Army drilled through the entire Greenland ice sheet down to bedrock at a site called Camp Century. They recovered 1,390 metres of ice core. Camp Century was a Cold War research base just beyond 77°N in the far northwest of Greenland, operated by the US military from 1959 to 1967. The ice-drilling programme sprang from a top-secret plan by the United States military called Project Iceworm to excavate 4,000 kilometres of tunnels within the Greenland ice sheet to house launch sites for nuclear missiles. The nuclear weapons project was never realised, but it stimulated a research programme to understand the dynamics of the Greenland ice sheet.

Dansgaard measured oxygen isotopes in ice samples throughout the long Camp Century core and reconstructed the temperature record of Arctic climate throughout the last glacial period. He published a landmark paper in the journal *Science* in 1969 called 'One Thousand Centuries of Climatic Record from Camp Century on the Greenland Ice Sheet'. This paper launched the new discipline of ice core science. It concluded that not only did ice core data provide far greater detail on past climate than any other available geological archive, but also that these data were continuous, with the potential to span several hundred millennia.

Ice cores from the Greenland ice sheet show distinctive annual layers produced by seasonal variations in snowfall, temperature and atmospheric conditions. The snow that falls in summer under the twenty-four-hour Arctic sun has a coarser texture and lower density compared to the snow that falls in the permanent darkness of the much colder Arctic winter. In some settings these paired layers can be counted like tree rings, but this ice stratigraphy becomes difficult to see in deeper, more compacted ice. Most layer counting is done with water isotopes that vary in concentration from summer to winter, giving us a high-resolution year-by-year dating framework for the climate history retrieved from the ice core. Remarkably, ice cores can provide an annual weather report for as many years as the record allows.

Perhaps the most significant breakthrough in ice core science came when Swiss and French scientists led by Hans Oeschger (1927–1998) and Claude Lorius (1932–2023) developed methods to measure carbon dioxide and methane gas concentrations in the air bubbles trapped in ancient glacial ice. This meant that the changing strength of Earth's greenhouse could be determined from ice core records. Bubbles of air trapped in the ancient ice tell us how the composition of the atmosphere has changed over time. To retain this important information, ice cores, once collected, must be stored in their pristine frozen state. After transport to a specialist laboratory, the cores are cut into thin slices; each disc of ice is either melted or crushed in a vacuum to release the ancient air so that its composition can be analysed.

Carbon dioxide is the most important planet-warming gas in our atmosphere. Almost every day we hear news about concentrations of carbon dioxide in the atmosphere being far in excess of *pre-industrial* levels. We only know what those pre-industrial levels are because of the analysis of air bubbles in ice cores, as continuous monitoring of Earth's atmosphere only began in 1958 at the Mauna Loa Observatory on Hawaii. Ice core scientists can measure changes in $CO_2$ through multiple glacial and interglacial cycles. The ice cores reveal the state of pristine Earth, providing the crucial baseline data from which we can measure the strengthening of Earth's greenhouse by the burning of fossil fuels.

Several ice cores more than 3 kilometres long have been recovered from the Greenland ice sheet. Between 1989 and 1993 the Greenland Ice Core Project (GRIP) drilled 3,029 metres from the Summit Camp down to bedrock at the base of the ice sheet while a project known as GISP2 (Greenland Ice Sheet Project 2) drilled to a depth of 3,085 metres from the summit of the ice sheet down to bedrock. These long ice cores have provided remarkably detailed records of environmental change for the last 100,000 years or so. Climate data retrieved from these ice cores show that the last glacial period was punctuated by repeated and rapid swings in temperature. These are known as Dansgaard–Oeschger Cycles.

Greenland's ice cores do not go so far back in time as those from Antarctica. Snowfall is much higher in Greenland than it is on the southern continent, so the ice builds up more quickly but represents less time. And very old ice does not have time to accumulate in Greenland because the ice sheet is more dynamic – ice flows to the coast more quickly. In contrast, the deep interior of Antarctica is a cold desert and one of the driest places on Earth. Antarctica has the thickest and oldest ice on the planet. Snowfall is extremely low – sometimes as little as one to two centimetres per year – so the ice sheet builds up only very slowly. The layers of ice are thinner, but the overall record is much longer.

In December 2004 the European Project for Ice Coring in Antarctica (EPICA) completed a drilling programme that yielded a core from Dome C on the East Antarctic ice sheet that reached a depth of 3,260 metres. The base of the EPICA ice core is dated to 800,000 years ago and the record spans the last eight glacial cycles of the Quaternary Period. The EPICA record revealed that the typical carbon dioxide concentration for an interglacial is 280 ppm (the pre-industrial $CO_2$ levels in the Holocene) and for a glacial period is much less, about 180 ppm. This record confirmed that the greenhouse effect was a key driver of ice age climate change. This EPICA core produced the most important insight from the ice core record – the revelation that global temperature and carbon dioxide rise and fall in lockstep through the glacials and interglacials of the Quaternary ice age. The EPICA record also showed that the concentrations of $CO_2$ and methane in today's atmosphere are completely unprecedented in the last 800,000 years.

Ice core research is not restricted to the big ice sheets in the polar regions. It is also possible to drill ice cores from glaciers in the highest mountains of the tropics, including the ice fields of the South American Andes and on Kilimanjaro in Africa. The climate records from tropical glaciers have shed light on the history of the monsoon and El Niño systems that dominate the climate of the tropical Pacific and affect global-scale oceanic and atmospheric circulation patterns.

Taking ice cores from some of the world's highest mountains poses considerable logistical challenges. This work has been pioneered by Lonnie Thompson (b. 1948) of Ohio State University, using light-weight solar-powered drilling equipment. When Thompson and his team cored glaciers in the Himalayas, yaks were used to carry the insulated boxes of ice cores down to elevations where freezer trucks could take over.

Quelccaya Ice Cap in the Peruvian Andes is the second-largest body of glacial ice in the tropics. Parts of it are 200 metres thick. Quelccaya ice cores contain thin layers of red windblown dust, deposited each year during the dry season. The distinctive dust layers can be counted to date the ice core record. Thompson's work on Andean ice cores has shown that they record alternating periods of drought and more humid conditions related to the El Niño Oscillation in the Pacific. They also contain ash layers that form a record of regional volcanic eruptions, as well as fragments of windblown charcoal that provide a long-term fire history for the Amazon rainforest.

While the environmental records from tropical mountain glaciers are much shorter than those from the Greenland and Antarctic ice sheets, some do extend back into the Pleistocene ice age. One record from the summit of Sajama mountain in Bolivia reaches back some 25,000 years. Even though all tropical glaciers have been in rapid retreat for several decades, Thompson and his team have collected over 7 kilometres of ice core, which remain in storage at Ohio State University. They have ice cores from sixteen countries, including ice from some tropical glaciers that no longer exist.

Ice cores provide unique archives of past environmental conditions on Earth. They are one of the most important sources of information on the history of the Earth over the past million years or so and work is in progress to push the ice core records from Antarctica even further back in time. Ice cores drilled from ice sheets provide the best long records of Earth's changing atmospheric conditions. As well as an unbroken record of climate change, they also form a record of the key mechanisms, such as the strength of the greenhouse effect, that force the climate to change.

# CHAPTER 34

# A Green Sahara

The German explorer and geographer Heinrich Barth (1821–1865) crossed the Sahara in 1850, travelling from Tripoli to Timbuktu. When he reached the Wadi Mathendous, one of the large dry riverbeds on the southern edge of the Messak Plateau in southwest Libya, he encountered an astonishing sight: spectacular life-size depictions of elephants, giraffes and crocodiles etched into smooth sandblasted rock faces. There were carvings of ostriches and hippos alongside domesticated animals, and everyday scenes of hunting, animal herding, dancing and music-making. Barth was one of the first Europeans to witness the Wadi Mathendous rock art, which is some of the oldest in the Sahara, dating from the early Holocene some 10,000 years ago.

North Africa has been described as the largest art museum in the world – Libya and Algeria boast an extraordinarily rich body of rock art in the heart of the Sahara Desert. On the dissected sandstone plateau of Tassili n'Ajjer in southeast Algeria, some 15,000 engravings have been identified dating from 11,000 years ago to recent centuries. In the baking heat of the desert these images record the changing physical, biological and cultural environments of the Holocene Epoch in North Africa. The rock art even records the introduction of the camel in the Sahara in the first millennium AD.

The images that Barth observed were far removed from the inhospitable, depopulated, bone-dry desert he was traversing. Several years later, in his *Travels and Discoveries in North and Central Africa* (1857), Barth wrote that the rock art 'bears testimony to a state of life very different from that which we are accustomed to see now in these regions'. He was contemplating a time in the not-so-distant past when the greatest desert on Earth was a land of plenty.

For the first half of the Holocene, between about 11,000 and 5,000 years ago, the largest hot desert on Earth did not exist. What is now the barren, hyper-arid Sahara was a vastly different place – it was a savannah environment that sustained large mammals like giraffes, buffalo and antelope, rivers flowed year-round, and permanent lakes, deep enough for hippos to swim in, dotted the landscape. The Sahara was green and blue: summer rains fell across North Africa because the monsoon extended its range as far north as the Mediterranean. This interval of very recent Earth history is known as the African Humid Period (AHP). It supported rich wetland ecosystems that drew animals and people into the deep interior of North Africa. These people created the remarkably rich body of rock art depicting the animals they saw in the landscape.

While the Quaternary Period is best known for the growth and retreat of ice sheets in the high-mid latitudes, the history of the Sahara shows that environmental changes in the tropics were just as profound. The African Humid Period of the Holocene Epoch was not a one-off – wet and dry cycles have punctuated the history of the Sahara deep into the Pliocene for at least the last 5 million years. Much earlier green Sahara periods allowed early hominins from East Africa to reach Morocco and Lake Chad. The sediments that accumulate on the floor of the Mediterranean provide a detailed archive of these environmental shifts because when the Sahara is wet and green, rivers transport sediments and organic matter from North Africa to the Mediterranean; when the desert returns, these rivers dry up and the Mediterranean receives only windblown dust.

Today the Sahara makes up one-third of the African continent, stretching from the Atlantic Ocean to the Red Sea. Its dryland terrains include bedrock plateaus, ancient lake beds, gravel-strewn plains, dry river channels known as *wadis*, mountain ranges with extinct volcanoes, and giant sand seas known as *ergs*, with dunes of various forms. Despite the popular image of deserts, dunes account for only a fifth of the Sahara. The desert is bounded to the south by a belt of savannah grassland known as the Sahel; to the north is the Mediterranean Sea.

Lake Chad lies on the southern edge of the Sahara, covering an area of 22,600 square kilometres straddling parts of Chad, Nigeria, Niger and Cameroon. Rarely more than several metres deep, this shallow tropical lake is sensitive to fluctuations in rainfall. During the Sahel droughts of the early 1970s, it shrank dramatically and fell below 2,000 square kilometres in area during the 1980s. At the peak of the AHP some 11,000 to 9,000 years ago, Lake Chad was 160 metres deep and covered an area of at least 350,000 square kilometres, which is about the size of Zimbabwe. The vastly expanded lake is known as palaeolake Megachad.

During the African Humid Period, palaeolake Megachad was the largest freshwater lake in Africa and may well have been the largest pluvial lake on Earth. A pluvial lake is one that has seen major changes in lake level and volume in response to climate change. The expanded Lake Chad filled a huge landscape feature known as the Bodélé Depression, which is bordered by ancient lake shorelines that can be mapped from space. Today the dried-out Bodélé Depression is the largest single source of atmospheric dust on Earth because the fine-grained lake sediments deposited during the AHP are now exposed to strong desert winds. When easterly winds are strong enough, this dust can cross the Atlantic Ocean and supply nutrients to the Amazon rainforest.

The Tibesti mountain range in the central Sahara is a zone of ancient volcanic activity, where numerous peaks and extinct craters lie above 2,000 metres. It is one of the most remote and least explored parts of Africa. The summit of Emi Koussi, at 3,415 metres above sea level, is the highest point in the Sahara. Emi Koussi is an ancient shield volcano with a base diameter over 70 kilometres across. The Era Kohor caldera forms part of the summit and is 2 kilometres in diameter and 300 metres deep. During the African Humid Period, along with many other calderas in the Tibesti Mountains, it was filled to the brim with water. These mountains provided a diverse high-level aquatic ecosystem for many species of birds, insects, fish, reptiles and mammals in the early to mid-Holocene.

To create permanent freshwater lakes in North Africa, annual rainfall inputs must exceed losses from evaporation. The evidence from across North Africa shows that, even in the most arid parts of today's Sahara Desert, there was an abundant supply of moisture to sustain these water bodies. Shells and other fossils embedded in lake shorelines have been dated by radiocarbon to show higher lake levels across the region about 9,000 years ago. The fossil record from pollen, seeds and preserved wood fragments shows that tropical trees and shrubs flourished some 500 kilometres north of their present distributions, typically as part of woodland communities fringing the rivers and lakes.

If you study satellite images of the Sahara you will see thousands of ancient river channels. Large rivers flowed north to the Mediterranean during the AHP, others flowed westwards to the Atlantic, many flowed eastwards to join the Nile. The monsoon rains filled sandstone and limestone aquifers with groundwater so that oases were plentiful, and flows were maintained in many rivers throughout the winter dry season. The perennial lakes and rivers of the AHP were a focus for human groups with herds of goats and cattle. The African Humid Period ended midway through the Holocene about 5,000 years ago. This phase of wetter climate ran counter to a long-standing and now defunct theory that glacial stages in lower latitudes were wet and interglacials were dry.

Today only one river flows in the Sahara. The Nile is the longest river on Earth. It is an *exotic* river, which means it originates in a wet region and then flows through a desert. The desert Nile begins at Khartoum in Sudan where the Blue Nile and White Nile meet. The Blue Nile rises in the mountains of Ethiopia, where monsoon rains generate the famous Nile flood that watered the crops of ancient Egypt. During the AHP the Nile flood was bigger than today, so more channels were needed to convey the enhanced flows.

Between 1995 and 2014 I spent nine winter field seasons in the Nile Valley of northern Sudan, collaborating with archaeologists from the British Museum. We found that the River Nile had a greater number of active channels during the African Humid Period because it carried higher flows from the more intense monsoon rains. In an 80-kilometre reach of the valley floor centred on the modern town of Dongola, the archaeologists mapped all the ancient settlements east of the main Nile and found that most sites that dated to the AHP were located *away* from Nile channels because they could be sustained by local rainfall. When the AHP ended and the Sahara became dry, all the archaeological settlements younger than 4,500 years old had shifted to the banks of Nile channels because the river was the only source of water in the desert. Further downstream in Egypt this aridification created a reliance on the waters of the Nile, thus sowing the seeds for the rise of the world's greatest riverine civilisation.

Why was the climate wetter across the Sahara between 11,000 and 5,000 years ago? To answer this question we need to look at long-term changes in the amount of solar energy received in the tropics. It is well known that differences in temperature between land and ocean drive monsoon circulation in the sub-tropics. The land heats up more quickly than the ocean: as warm air rises over the land it creates a large zone of low pressure that draws in moist air from the ocean. This is known as a monsoonal flow. The reverse happens in winter, producing a long dry season. The monsoon was stronger during the AHP, 10,000 years ago, because North Africa received about 7 per cent more solar warming during the summer than today and this extra energy produced a greater difference between the temperatures of land and ocean. This generated a stronger monsoonal flow.

This increase in monsoon strength was caused by the *precession cycle*, which is one of the astronomical cycles studied by Milutin Milankovitch that drive long-term change in Earth's climate (Chapter 32). The precession cycle has a particularly strong influence on rainfall in the tropics because it changes the season when Earth is closer to the Sun. Today, Earth is closest to the Sun (*perihelion*) in January, but 10,000 years ago (roughly half a 23,000-year precession cycle) it was closest in June or July, and this produced a 7 per cent increase in summer radiation. This might not seem like very much, but it was enough to create a larger contrast in temperature between land and ocean, which strengthened the monsoon across North Africa.

The monsoon is more intense (with more rainfall) and has a greater geographical extent when the precession cycle puts perihelion in the northern hemisphere's summer and therefore increases the amount of summer radiation. When summer insolation decreases, the monsoon becomes less intense and its geographical range contracts. Over tens of thousands of years both the intensity and reach of the African monsoon shift in tandem with these orbital changes.

The Sahara as we know it is less than 5,000 years old. It has alternated between humid and dry for at least 5 million years as the monsoon has waxed and waned at the command of the Earth's wobbly rotation. By studying climate history in deserts we can improve our understanding of the global climate system and get a better idea of what changes might be in store as the Earth continues to warm. It remains to be seen whether the future of the Sahara will be green.

# CHAPTER 35

# Taming the Earth

For more than 95 per cent of human history, *Homo sapiens* sourced their food by foraging, hunting and scavenging. Foraging involved gathering a wide range of foodstuffs from edible wild plants and from rivers, lakes and coasts. By the tail end of the Pleistocene some 12,000 years ago, human groups began to shift from a lifestyle of mobility and temporary camps to one based around more permanent settlements. Plants and animals were domesticated as societies transitioned to selective hunting, crop planting and livestock herding, a process of change that played out over many generations. This period in human history is called the Neolithic (New Stone Age) because stone remained a key part of the human toolkit. The shift to farming in the Old World is often called the Neolithic Revolution.

It is difficult to overstate the significance of this cultural transformation – it ranks alongside walking on two legs and the use of fire in the human story. We know from archaeological excavations and radiocarbon dating that agriculture arose at different times in different parts of the world, but we don't know *why* the shift to agriculture took place. This is one of the most intriguing and fiercely debated questions in human prehistory. Some archaeologists have argued that farming was enabled by the more stable climate of the Holocene Epoch and, because humans had spread to occupy almost all the habitable areas of the Earth, a better way to feed growing populations was needed. The development of agriculture may have allowed humans more control over their food resources and nutrition as well as offering a more stable lifestyle with a degree of protection against climatic uncertainty. We cannot be certain about the motivations behind this profound change in lifestyle, but we do know that it utterly reshaped our relationship with the Earth.

Vast areas of the Earth's surface have been transformed by agriculture. Forests have been cleared, soils ploughed, hillslopes terraced, paddy fields irrigated, plants and animals have been domesticated and selectively bred. These processes have radically changed soils, landscapes, watercourses and biodiversity on every continent apart from Antarctica. Today about 37 per cent of land is used for food production: approximately 11 per cent for growing crops and about 26 per cent for grazing livestock. This includes some 1.5 billion head of cattle and about 1.3 billion sheep.

The practice of agriculture was closely linked to the formation of settled communities leading to specialist labour – shepherds, pot makers, weavers, carpenters, millers – and the creation of marketplaces. As population increased and technology advanced, the first cities and complex societies arose in the ancient world. A reliable supply chain of foodstuffs from field to town was essential to sustain this new way of living.

A geographical area known as the Fertile Crescent that includes parts of present-day southeastern Turkey, Syria, Iran, Iraq, Jordan, Palestine and Israel is considered the birthplace of agriculture. This region includes lands close to the eastern Mediterranean, and a thick belt of land tracking the course of the Tigris and Euphrates rivers to the Persian Gulf. This part of southwest Asia contains the earliest evidence of plant and animal domestication dating to between 12,000 and 11,000 years ago, close to the Pleistocene–Holocene boundary. Sheep, goats and cattle were domesticated within the Fertile Crescent, alongside the cultivation of crops such as wheat, barley, lentils and chickpeas. This is where the earliest settled farming communities emerged by around 10,000 years ago.

The origins of settled life are being pushed back further in time as new sites are excavated and dated. Göbekli Tepe is one of several remarkable sites that have added new details to this story. Located in the foothills of the Taurus Mountains in the headwaters of the Euphrates river, this settlement was occupied from around 11,500 to 10,000 years ago by a village-based community of hunter-gatherers. Excavations have uncovered grinding stones indicative of cereal processing. The site is most famous for its monumental megalithic structures over 5 metres tall and engraved with images of wild animals including scorpions, lions, foxes and vultures. It has been dubbed the world's oldest temple, since the megaliths predate those at Stonehenge by at least 6,000 years. Whether or not this site was a temple is open to debate; its T-shaped pillars carved from local limestone supported large communal buildings. Göbekli Tepe was one of the world's first settled communities with astonishingly rich evidence of innovative building technology.

Domesticated plants have been modified profoundly by thousands of years of evolution under cultivation. Wheat was one of the first plants to be domesticated and remains one of the most important crops today, accounting for about one-fifth of calorie intake. There are key differences between the wild wheat ancestor of the Fertile Crescent and domesticated varieties. The latter tend to be taller, with larger seeds and a more robust stem; they also flower later, which increases yield and helps with harvesting. Genetic studies are revealing new insights into the domestication process and the pace of change – modern wheats have been shaped by more than 10,000 years of human-influenced evolution. Features of domestication emerged via unconscious selection in ancient times as well as more recently by deliberate selective breeding.

Agriculture spread westwards from the Fertile Crescent across the Mediterranean and into the rest of Europe, reaching Britain by about 5,500 years ago. Cyprus is the closest Mediterranean island to the Fertile Crescent and was part of the early farming revolution. There is well-dated evidence for the presence of domesticated animals on Cyprus in the very early Holocene, which sheds light on the capabilities of early Holocene Mediterranean seafarers. Neolithic people were transporting young cows on wooden boats from mainland Anatolia to Cyprus 10,500 years ago. These voyages could have been made several times a year in craft that were large enough for weaned calves to stand during the journey.

While hunting and gathering gave humans an intimate understanding of animals and plants and their habitats, domestication involved animals and plants becoming *dependent* on humans for their well-being and survival in exchange for various services. Cattle, goats and sheep provided milk and dairy products as well as meat. Cattle also provided power to drag ploughs and cultivate larger and steeper areas. The earliest plough was the ard, a lightweight wooden tool that cuts a shallow furrow for planting seeds but does not turn over the soil. Its origins are uncertain, but ards made from iron have been discovered in Mesopotamia (4,300 years ago) and the Nile Valley (4,500 years ago). Excavations of Neolithic sites occasionally yield evidence of plough use through the preservation of furrows in buried soils. Agricultural scenes are common in Old Kingdom and later Egyptian pyramid tomb art, showing ploughing and harvesting on the Nile floodplain.

Radiocarbon dating from archaeological sites across the Fertile Crescent shows that settlements like Göbekli Tepe supported by agriculture were widespread by 10,000 years ago. Today much of the lowlands across this region are semi-arid but the early Holocene climate was, like the green Sahara, wetter than today, allowing the cultivation of a range of crops without the need for irrigation. However, ever since cultivation began in southern Mesopotamia, its farmers have grappled with the problem of salts building up in floodplain soils under a hot desert climate. When irrigation waters evaporate, sodium and magnesium salts precipitate, leaving a white crust on the soil surface. Too much salt hinders seed germination and limits the uptake of water and nutrients by plants. Some Mesopotamian farmers planted barley instead of wheat because it copes better with an increase in salinity, but when yields fell so that farming was no longer viable, areas were abandoned.

The Neolithic Revolution in China is well documented in the archaeological records of its two mighty rivers. The humid sub-tropical climate of the Yangtze river valley was well suited to rice cultivation in flooded paddies, where early farming communities built dwellings on stilts within floodplain wetlands. Farming began in the Yellow (Huang He) river valley about 9,000 years ago, with various types of millet that grow well in the drier and cooler conditions of northern China. The emergence of agriculture on these two great floodplains was central to the rise of Chinese civilisation. Rice and millet became the staple foods of the south and north respectively, shaping later regional cultures. Pigs and chickens were the most important domesticated animals, along with water buffalo. The domesticated crops and agricultural practices from these early farmers in China eventually spread to other parts of East and Southeast Asia. Millet and various pulses were domesticated in India by 5,000 years ago.

In the Americas the potato was domesticated in the Andes by 7,000 years ago and quinoa by about 5,000 years ago. The potato was a staple of the Inca civilisation for centuries and was brought to Europe by Spanish conquistadors in the sixteenth century. Maize was first cultivated in Central America around 9,000 years ago and the squash may have been earlier, underpinning the rise of civilisations including the Maya and Aztecs.

The spread of agriculture in the Holocene provided food security for the growth of towns and cities and complex human societies. For much of the last century it was widely believed that humans only began to substantially modify their environment when farming began. There is no doubt that ploughing soils and the domestication of plants and animals had a profound impact on landscapes and ecosystems. Clearing woodland and ploughing both increased soil erosion and the amount of fine sediment washed into rivers. But there has been much debate about when humans first began *deliberately* to modify the environment for their own benefit. A growing body of evidence is emerging pointing to a deeper history of deliberate modification of ecosystems, even if this may have been on a more local scale.

For instance, the onset and extent of deliberate burning by Aboriginal people in Australia has been much debated. Humans first arrived in Australia between 65,000 and 50,000 years ago and used cultural burning for much of that time. One of the main challenges with working out the extent to which humans used fire in the past to manage ecosystems is how to distinguish between natural and human-caused wildfires in the sedimentary record. This is not straightforward when an increase in the concentration of charcoal fragments is the main evidence.

Deliberate ecosystem modification by humans in Europe goes back at least as far as the Neanderthals. The evidence is not always clear cut, but some archaeologists have argued that Neanderthals used fire to burn woodland on the edge of lakes to keep the landscape open, thereby creating a wider range of plant resources for them to exploit. These clearings would also attract large herbivores, which were hunted and butchered. There is evidence from the ancient lake basin of Neumark-Nord in northern Germany that 125,000 years ago Neanderthals maintained an open landscape for about 2,000 years in a lake margin setting that would otherwise have been forested. This ecosystem supported a rich fauna of straight-tusked elephant, rhinoceros, wild boar, horse and large cattle, as well as several species of deer, lion, hyena and bear. Excavations have revealed thousands of stone tools with cutmark-scored animal bones alongside huge amounts of charcoal.

Domestication has deep roots too. All breeds of dog alive today are descended from the grey wolf (*Canis lupus*), the domestication of which took place in the last glacial period, between about 40,000 and 20,000 years ago when humans were hunter-gatherers. Dogs were the first domesticated animals, and the first to form a close, cooperative relationship with humans, providing important benefits through hunting, companionship and protection. In the last glacial period, wolves that were less aggressive and more tolerant of humans may have scavenged near human camps for scraps of food. In turn, humans encouraged this relationship by feeding the less aggressive wolves so that natural selection led to the dominance of traits that favoured a mutually beneficial relationship and, over time, dogs behaved very differently to their wild ancestors. Man's best friend goes back a long way.

# CHAPTER 36

# The Little Ice Age

Between 1608 and 1814 the River Thames in central London froze over twenty-four times. In the coldest winters the ice was as thick as a coffin and stayed frozen for several weeks. All water trade and transport stopped and Londoners set up carnivals on the ice-bound river. These became known as frost fairs. In the great frost of 1683–4, when the Thames was frozen from bank to bank for two months, a remarkable winter scene ensued, with fox hunts, hog roasts, bowling tournaments and printing presses on the river. There were drinking tents and rows of stalls decked in colourful flags and banners, surrounded by scores of people skating on the ice. Even the king joined the fun: a printed frost fair souvenir for 1684 bears Charles II's name. The last fair in 1814 was the shortest, lasting barely a week, but the ice was strong enough to take the weight of an elephant at Blackfriars Bridge. Thanks to the artists who painted these scenes, the frost fairs on the Thames became emblematic of a period known as the Little Ice Age.

The Little Ice Age was a 600-year period of cooler-than-average temperatures, from about 1250 to the middle of the nineteenth century. It followed an interval of warmer conditions from about 900 to 1250, known as the Medieval Warm Period. There has been much debate about the magnitude of the cooling during the Little Ice Age and whether it was experienced in all parts of the globe. The impacts were certainly not uniform in extent or intensity, but they are most clearly expressed in the northern hemisphere. It was just before the Second World War that Dutch-born geologist and gifted map maker François Matthes (1874–1948) first coined the term Little Ice Age, following fieldwork in California's Sierra Nevada that showed its glaciers had been more extensive in the very recent past.

In 1988 the British physical geographer Jean Grove (1927–2001) published the seminal book on this period based on three decades of research. *The Little Ice Age* brought together data on past climate from tree rings, peat bogs and ice cores and combined them with the results of her own fieldwork in Norway and the Alps. Grove was a glaciologist who could read the landscapes of ice-scoured Alpine valleys. She reconstructed the history of glacier fluctuations and worked out their climatic significance. Jean Grove found the same pattern almost everywhere she looked: Alpine glaciers had advanced for several centuries before retreating after 1850. She also extracted information on past climate from historical documents such as diary entries on crop failures and sea ice sightings from North Atlantic ships. Her work is important because she was able to demonstrate from an eclectic mix of sources that the climate of the last 1,000 years had not been constant.

For much of the twentieth century a dominant view in Earth science and physical geography held that the Holocene climate had been stable – it was regarded as uneventful, even rather dull. While the great climate swings of the Pleistocene ice age were clear for all to see, there was an especially stubborn orthodoxy resistant to the idea of climate change during the historical period. Jean Grove was one of a tiny band of academics who ignored these climate change sceptics and became intrigued by historical changes in climate long before the topic became fashionable.

British climatologist Hubert Lamb (1913–1997) played a key role in making the case that climate *had* changed in the historical period. The climate changes of recent decades are such a prominent part of modern discourse that it is hard to believe that this topic was so far from the mainstream as recently as the early 1970s. Lamb was a pioneer of historical climatology who, like Jean Grove, could tease out the big picture from scattered fragments of information. He explored past variations in climate and demonstrated not only that the climate changed over human timescales, but that those changes were large enough to impact human affairs. Lamb's innovative use of written sources was often sniffed at by influential figures in the climatology and wider Earth science establishment, who saw his approach as descriptive and non-scientific. Grove and Lamb argued forcefully that knowledge of historical climatic variations was essential to fully appreciate the nature and impacts of future climate change.

There are many intriguing stories that point to a cooler climate in northwest Europe during the Little Ice Age. Lamb describes how between about 1690 and 1728 there were several reports of Inuit kayaking to the Orkney islands off the northeast tip of Scotland. There was one account of a kayak so far off course it reached the mouth of the River Don near Aberdeen. Lamb attributed these adventures to the southward expansion of Arctic waters and sea ice. In the coldest part of the Little Ice Age the ocean surface in parts of the North Atlantic may have been 4 to 5°C cooler than in the late twentieth century. This cooling explains the documented decline of the cod fishery in the seventeenth century, since cod cannot survive in such chilly waters.

John Ruskin (1819–1900) travelled widely in the Alps in the nineteenth century, taking a keen interest in glaciers and glacial processes. In a diary entry for 10 July 1835 he marvelled at the drama and dynamism of the Glacier des Bossons, which flowed all the way down the north side of the Mont Blanc massif to the floor of the Chamonix valley: 'It is far superior in beauty to the Mer de Glace, broken with splendid pyramids of dazzling ice – its moraine however is very large, and all moraines are very ugly, for owing to the constant motion of the glacier the moraine is kept in motion too.' Two centuries before, in the middle of the Little Ice Age, the Glacier des Bossons had advanced so far across farmland and houses in the village of La Fouly that the bishop was summoned to rid the glacier of its demons. When Ruskin returned to the Alps in the summer of 1874 he saw glaciers in retreat: 'I was able to cross the dry bed of a glacier, which I had seen flowing, two hundred feet deep, over the same spot, forty years ago.' When I visited the Chamonix valley in 2015, the Glacier des Bossons was a shadow of its former glory, having thinned markedly and receded several kilometres upslope.

The Little Ice Age may also explain the mysterious disappearance of the Norse settlements on Greenland. In AD 985, during the Medieval Warm Period, the great Norse explorer Erik the Red led an expedition to Greenland, paving the way for the establishment of permanent settlements. At its peak the Norse population exceeded 3,000. But sometime in the fifteenth century the Norse disappeared. Did they impose livestock and farming methods that were unsuited to the Arctic? Did they fail to adapt to the climate changes of the Little Ice Age? The demise of the Greenland Norse is one of the great mysteries of Arctic archaeology.

The narrative on the Norse demise has shifted in recent decades as archaeological excavations have pointed to a community that was more reliant on food from the sea than it was on farming and rearing livestock. The Norse also exploited the marine ecosystem for walrus ivory, which was highly prized in Europe. It seems unlikely that these exceedingly resourceful people failed to sustain their livelihoods on Greenland because they were stubborn incompetent farmers. The picture was more complex, since it is likely they were hunters who farmed, rather than the other way around.

The Little Ice Age climate no doubt made life tougher on Greenland, but it also disrupted trade as sea ice expanded and the North Atlantic became stormier. As elephant ivory from Africa appeared in the European markets, the falling value of walrus ivory certainly did not help. A shorter growing season and harsher winters led to the abandonment of farms as the Little Ice Age began to bite, but it may have been a labour shortage that made settlements unviable. Ironically, much of the archaeology that tells this story is currently under threat as organic materials left by the Norse deteriorate in a warming climate.

Several explanations have been put forward to account for the cooler climate of the Little Ice Age but there is no real consensus. One idea mooted was reduced solar activity in the period between 1645 and 1715. This is the Maunder Minimum, which takes its name from the astronomer Edward Maunder (1851–1928), who spent many years at the Royal Observatory in Greenwich photographing and measuring sunspots. But this period of reduced solar activity cannot be the main cause behind the Little Ice Age because the climate became cooler *before* the solar minimum began. It may well have contributed to the cooling, but it did not trigger the Little Ice Age.

There were big volcanic eruptions in various parts of the world during the Little Ice Age that blasted sun-blocking aerosols high into the atmosphere. One of the largest climate-modifying volcanic eruptions of the past thousand years took place in Indonesia on the island of Lombok in 1257, right at the beginning of the Little Ice Age. The eruption column is estimated to have reached altitudes of 43 kilometres. It was followed in 1258 and 1259 by two of the coldest northern hemisphere summers of the past thousand years. The notorious Lakagígar (Laki) fissure eruption in Iceland of 1783 lasted eight months, producing a volcanic haze that extended as far as Syria. Crops failed and half of Iceland's cattle died from eating fluorine-contaminated grass. This eruption was followed by a very severe winter across Europe in 1783–4, with bitterly cold temperatures, frozen fields, thick snow cover and ice-bound lakes and rivers. When the snow melted there was widespread flooding north of the Alps. The 1815 eruption of Tambora in Indonesia was the most powerful in recorded history, releasing enormous amounts of ash and gases. It produced a few years of global cooling, including the year without a summer in 1816. While these eruptions helped to sustain Little Ice Age cooling, it is difficult to attribute the long centuries of cooler climate to a purely volcanic cause, whose climate impacts were mostly short-lived.

The most radical theory suggests that diseases such as smallpox, influenza and measles brought to the Americas following European colonisation decimated Indigenous populations and led to the collapse of their labour-intensive farming systems. The regrowth of natural forests was so extensive throughout the Americas that it trapped enough carbon dioxide to weaken the greenhouse effect, albeit the impact on global climate was very minor.

The geography of Little Ice Age cooling is complex and associated with quite small changes in average temperature. While average annual temperatures in the northern hemisphere only fell by about 0.6°C, some regions saw temperatures fall by as much as 2°C relative to thousand-year averages. But even small changes in temperature have consequences for glaciers, ecosystems and people. While the interpretation of historical sources is not always straightforward, the records compiled by Lamb, Grove and others show that Earth's climate system is finely tuned.

We now have robust high-resolution and well-dated proxy climate data for all 11,700 years of the Holocene. These records provide important context for the warming of recent decades. We will look at some of these records in the next chapter.

# CHAPTER 37

# Warming the Earth

As man is now changing the composition of the atmosphere at a rate which must be very exceptional on the geological timescale, it is natural to seek for the probable effects of such a change.

If you said this today, almost no one would bat an eyelid. Predicting future scenarios in a world warmed by human impact is quite rightly the focus of many working in and reporting on climate science. But this statement was published back in 1938, some five decades before the international scientific community began to accept the reality of global warming driven by human activity. It was the remarkably prescient view of a steam engineer, inventor and amateur scientist who concluded, from an analysis of decades of thermometer measurements, that Earth was getting warmer and carbon dioxide produced by the burning of fossil fuels was the key driver of that warming. Although Guy Callendar (1898–1964) had pieced together the key elements of anthropogenic global warming, his ideas were largely forgotten.

Callendar studied temperature records from 147 weather stations across the world, poring over the data and making thousands of calculations by hand. His analysis showed that the average temperature of the land areas on Earth had increased by 0.3°C over the previous fifty years. This may not seem very much, but Callendar was the first person to establish from meteorological records that the Earth was warming. This was a critical milestone in modern climate science.

By gathering together the few scattered measurements on the concentration of carbon dioxide in the Earth's atmosphere, Callendar was also able to show that the abundance of this greenhouse gas was steadily rising, and he attributed this rise to the burning of coal. Callendar estimated that the burning of fossil fuels had added some 150 billion tons of $CO_2$ to the atmosphere in the previous five decades. Although Callendar's approach might be regarded as crude today, his main conclusions were sound. His pioneering insights that fossil-fuel burning increased atmospheric $CO_2$ and caused Earth to warm became known as the 'Callendar Effect', and have been fully vindicated. But his work received a chilly reception from his contemporaries and went unnoticed amid the chaos of the Second World War. Despite his fundamental contribution to Earth science, Callendar is still not widely known outside the climate science community.

After publishing his seminal paper in 1938, Callendar immediately encountered sceptics – as he knew he would. He anticipated such resistance in the opening lines of his paper, when he said that 'Few of those familiar with the natural heat exchanges of the atmosphere, which go into the making of our climates and weather, would be prepared to admit that the activities of man could have any influence upon phenomena of so vast a scale.' The scientific community had great difficulty believing that humanity could modify a planetary-scale energy budget. In his paper Callendar 'hope[d] to show that such influence is not only possible, but is actually occurring at the present time.' One who took up the challenge was a young American scientist called Charles David Keeling (1928–2005), who was inspired by Callendar to begin systematic measurements of the carbon dioxide concentrations in Earth's atmosphere.

Direct measurements of $CO_2$ in the Earth's atmosphere began in 1958 at the Mauna Loa Observatory in Hawaii. This is an ideal location from which to monitor the composition of the Earth's atmosphere. It is located in the central Pacific Ocean, over 4,000 kilometres from the Californian coast, and sits at an elevation of some 3,400 metres above sea level on the north side of Mauna Loa volcano, where the air is well mixed and a long way from major sources of atmospheric pollution in Asia and North America. The measurements taken in Hawaii represent average concentrations for the global atmosphere.

Daily recordings of carbon dioxide at the Mauna Loa observatory are maintained by the Scripps Institution of Oceanography at the University of California San Diego. Keeling maintained the measurements for some five decades until his death in 2005. This long-term monitoring effort was pivotal in showing that the use of fossil fuels across the world was changing the composition of the Earth's atmosphere and strengthening greenhouse warming. It is widely known as the Keeling Curve. It shows that carbon dioxide concentrations have increased year on year from the first measurement of 315 parts per million in 1958. They passed 350 ppm in 1987 and 400 ppm in 2014. The rate of increase has escalated every decade since monitoring began. At the beginning of 2025 the figure was 426 ppm, roughly 150 ppm above the pre-industrial interglacial level of 270 to 280 ppm that we know from ice core records (Chapter 33).

By studying the long-term data from the Mauna Loa Observatory, Keeling was the first person to recognise that the concentration of carbon dioxide in the Earth's atmosphere shows a clear seasonal pattern. Most of the Earth's land surfaces are in the northern hemisphere and this is where the major expanses of forests and other vegetation types are found that see pronounced seasonal variation. When vegetation begins to grow again in the northern hemisphere's spring, plants absorb carbon dioxide from the atmosphere via photosynthesis. In this way carbon is moved from the atmosphere to the terrestrial biosphere. This is a key element of the global carbon cycle, a process that takes place on such a grand scale that we see a fall in the concentration of carbon dioxide in the atmosphere from late spring to the end of summer (May to October).

When trees shed their leaves and plants die back during the northern hemisphere's autumn and winter, atmospheric concentration of carbon dioxide increases. Think of the spectacular colours in the deciduous woodlands of New England as the leaves turn during fall. As the leaves die and fall to the ground, the carbon in their tissues begins a new phase – carbon dioxide is released back into the atmosphere as microorganisms and fungi decompose the leaves. Some of the carbon is added to the soil's carbon pool as organic matter.

This seasonal pattern is superimposed upon the overall increasing trend of rising carbon dioxide concentrations resulting from the continuous addition of carbon dioxide from human activities. Each seasonal peak and trough is higher than that of the previous year. Processes such as forest fires and thawing Arctic permafrost can also add $CO_2$ to the atmosphere.

Charles David Keeling's groundbreaking research at the Mauna Loa Observatory has proved critical in understanding the role of greenhouse gases in planetary warming and in climate change. It has also helped us work out the causes of climate change in earlier geological periods. The measurements he started on top of a Hawaiian volcano in the middle of the Pacific Ocean continue to this day. They form one of the most important datasets in environmental science.

As we have seen, the oceans themselves play a crucial role in regulating Earth's climate system. They absorb 25–30 per cent of the carbon dioxide that human activities currently emit to the atmosphere. Since the beginning of the Industrial Revolution, roughly 500 billion tonnes of $CO_2$, or about 40 per cent of the total anthropogenic $CO_2$ emissions over that period, have been absorbed by the oceans. That has fundamentally changed their chemistry. Some $CO_2$ is taken up by phytoplankton during photosynthesis, and when these organisms die or are consumed, this carbon sinks to the deep ocean carbon store. Some $CO_2$ dissolves into the ocean's surface, lowering the pH of seawater and making it more acidic and more likely to dissolve shells and corals made from calcium carbonate. Ocean acidification can weaken coral resilience, making it more vulnerable to bleaching from heat stress and disease. The very life-forms in the ocean that help to sequester carbon by building shells and structures of carbonate are under attack from the carbon dioxide dissolved in seawater. And, as increasingly acidic oceans warm due to climate change, their capacity to absorb $CO_2$ decreases. The ocean is a critical buffer in the global carbon cycle, but its capacity to absorb $CO_2$ is limited, and the associated impacts like acidification pose significant risks to marine ecosystems.

In 1988, when the concentration of carbon dioxide in the atmosphere was 350 ppm, climate scientist James Hansen (b. 1941) delivered a historic testimony to the US Senate Committee on Energy and Natural Resources. Hansen was the director of NASA's Goddard Institute for Space Studies in New York and a much-respected figure. He declared that global warming was in progress and directly linked to the burning of fossil fuels. Hansen argued that the global temperature rise matched the patterns predicted by climate models and could not be explained by natural variability alone. He warned that unless action was taken to reduce greenhouse gas emissions, the Earth would face significant and potentially dangerous changes in climate.

James Hansen's testimony was a landmark moment in climate science communication, which raised political and public awareness of planetary warming. It also set out the urgent need to tackle the problem. His testimony came exactly fifty years after Guy Callendar's demonstration that Earth's temperature was rising because of human activities that began with the Industrial Revolution. Not only did it bring the issue of climate change into the political spotlight, it also marked the first time that climate change was recognised as a pressing policy issue at a national level. Hansen's testimony played a key role in the creation of the Intergovernmental Panel on Climate Change (IPCC) later that year.

Further fuel for the fire came from a landmark 1989 paper written by three climate scientists, showing temperature change in the northern hemisphere for the previous 1,000 years. Michael Mann (b. 1965), Raymond Bradley (b. 1948) and Malcolm Hughes (b. 1943) used tree rings and other 'proxy' evidence to reconstruct past temperatures. Their graph charting the recent abrupt upward trend in temperature became known as the *hockey stick* because it resembled the shaft and blade of an ice hockey stick. It soon attained iconic status in the climate change debate.

After some 850 years of modest natural climate variability that includes the Little Ice Age, the initial rise in temperatures coincides with the onset of the Industrial Revolution and increased greenhouse gas emissions, especially $CO_2$ from burning coal. The steep 'blade' of the hockey stick marks the period from the end of the Second World War when temperatures rise rapidly and deviate far above natural variability. The sharpest increase in temperature is from the 1970s onwards, driven by exponential growth in fossil fuel use and deforestation. The hockey stick graph triggered huge debate when it was published but the accuracy of the record has been reinforced by later work confirming the unprecedented nature of recent warming. The sharp upward trend has continued to the present day.

The United Nations Framework Convention on Climate Change (UNFCCC) Paris Agreement, adopted in December 2015, was signed by 195 countries. It is a global framework to combat climate change and accelerate actions for a sustainable low-carbon future. A key goal of the Paris Agreement was to place a limit on global warming to stop the Earth from warming 2°C above pre-industrial levels. This was the first time in history that an agreement had been struck to control the Earth's temperature. There was also an aspirational target to limit the temperature increase to 1.5°C above pre-industrial levels in order to significantly reduce the negative impacts of climate change on humanity and the living world. Unfortunately, that critical milestone has already been passed, when in 2024 global temperatures were more than 1.5°C above pre-industrial levels for the first time.

It is a sobering thought that we are modifying the carbon cycle and warming the planet more quickly than at any time since humans first walked the Earth. Earth history shows us that the warming of the planet over the next century will be controlled by the amount of carbon dioxide that human activities add to the atmosphere. But there is an alternative path. By speeding up the green energy transition from fossil fuels to renewable, cleaner, low-carbon energy sources, we can slow the rate of warming and leave a habitable world with cleaner air and greater energy security for future generations. It is up to us to try.

# CHAPTER 38

# Planetary Boundaries

In 1962 marine biologist and nature writer Rachel Carson (1907–1964) asked the world to contemplate a morning without birdsong. Her hugely influential book *Silent Spring* exposed the dangers of the most powerful chemical pesticide ever developed. DDT was capable of wiping out hundreds of species of insect in a single treatment, with devastating impacts on food webs and ecosystems. Despite fierce personal attacks by the chemical industry to undermine her work, Carson's book forced the US government to ban DDT and other pesticides. It also set in motion new thinking about the living world that led to the establishment of the US Environmental Protection Agency and the passing of laws to protect both the environment and human health. *Silent Spring* inspired the international environmental movement that began to mobilise in the 1960s. For the first time, Rachel Carson's groundbreaking book forced people to reckon with the fact that humans were capable of destroying the biodiversity that underpins the habitability of our planet.

We've seen in earlier chapters the various factors that combine to make Earth habitable. Earth is neither too hot nor too cold; its distance from the Sun provides the right amount of heat to allow liquid water to exist in abundance at the planetary surface. And water has remarkable properties – it is the ideal solvent for the chemical reactions needed to generate life. Earth has a fully functioning planetary-scale water cycle which is crucial for many processes, including bringing rainfall to the continents, moving heat around the planet, building ice sheets, returning water to the oceans and transporting minerals and nutrients.

Earth's atmosphere of mainly oxygen and nitrogen provides breathable air for complex lifeforms, while the ozone layer protects animals and plants from harmful ultraviolet radiation. Our world is kept warm by an insulating atmosphere – greenhouse gases prevent Earth from being stuck in a permanent ice age. The magnetic field is strong enough to shield both the atmosphere and planetary surface from cosmic radiation and geomagnetic storms. The churning metallic core deep in Earth's interior is the right size and composition to sustain this magnetic field.

Earth is the only known planet with active planetary-scale plate tectonics. Plate movements help to prevent extreme swings in Earth's temperature becoming the norm by regulating the geological carbon cycle through processes of uplift and weathering. This has helped to maintain, over billions of years, a climate suitable to sustain life. Life as we know it requires liquid water, energy and nutrients. Earth's mantle and volcanic processes add gases to the atmosphere and have supplied the key elements needed for life including carbon, hydrogen, oxygen, nitrogen, phosphorus and sulphur. Biological materials like DNA, proteins and carbohydrates are made up of these elements.

The existence of a large stable Moon for much of Earth history has helped to stabilise the tilt of the Earth and prevent extreme shifts in seasonality and climate. The tidal cycles controlled by the Moon may have played a role in the first colonisation of land by complex animals. Time is important too. The combined influence of the factors listed here has yielded billions of years of remarkable planetary stability, allowing life to evolve and diversify. Of course, Earth history has not been without drama, including planetary-scale catastrophes like Snowball Earth deep freezes, giant asteroid impacts and multiple mass extinctions. And in the most recent chapter of the story, human activity has now emerged as the latest threat to planetary habitability.

Ideas about the habitability of our world and the processes that produce conditions favourable for life have deep roots that extend to the very beginnings of geology as a scientific discipline. In 1788, in his seminal work *Theory of the Earth*, James Hutton talked about the 'globe of the earth' as 'a habitable world' dependent on 'the form of the whole, the materials of which it is composed, and the several powers which concur, counter-act, or balance one another'. In other words, even at the end of the eighteenth century, Hutton saw the Earth as an interconnected system with a set of large-scale processes (powers) that were responsible for creating a habitable world. Hutton was also a farmer who understood the land and was concerned about soil erosion; he appreciated that a deep knowledge of the workings of the Earth system was key to understanding this habitable state.

Hutton lived at a time of great change, as the Industrial Revolution gathered pace and Enlightenment ideas about reason, empiricism and progress circulated. He had long discussions with philosopher David Hume (1711–1776), economist Adam Smith (1723–1790) and chemist Joseph Black (1728–1799), who discovered carbon dioxide. With their commitment to interdisciplinary dialogue, I like to think this group would have engaged enthusiastically with a new way of thinking about how to maintain the habitability of our planet proposed by an international team of scientists in 2009. Against a background of global warming, habitat destruction and widespread pollution of the environment, they measured the human impact on Earth's environmental systems via nine *planetary boundaries*, which, in their view, define a safe operating environment for humanity. The planetary boundary framework identifies the environmental limits within which humanity can safely operate without causing catastrophic damage to Earth's life support systems.

The planetary boundary approach was conceived by Stockholm University's Johan Rockström (b. 1965) and a large group of international collaborators including James Hansen, the NASA scientist who put global warming on the political agenda in the 1980s, and Nobel Prize-winner Paul Crutzen (1933–2021), who has done most to popularise the concept of the Anthropocene (Chapter 40). Their 2009 paper published in *Nature* explored the *safe operating space* for the following nine planetary systems to ensure continued human and ecological well-being:

- Climate change
- Biodiversity loss
- Land use change
- Biogeochemical flows (nitrogen and phosphorus cycles)
- Fresh water use
- Ocean acidification
- Atmospheric aerosol loading
- Ozone depletion
- Novel entities (chemical pollution and plastics)

The 2009 analysis showed that the boundaries for three of these systems (climate change, biodiversity loss, and human interference with the nitrogen cycle) had already been exceeded. If Earth crosses too many boundaries, we risk leaving Holocene-like conditions – considered a safe zone for humanity – and creating a less hospitable world for future generations. By crossing key planetary boundaries such as climate change and biodiversity loss, we risk shifting Earth into an unfamiliar state such as those of the deep past that saw mass extinctions.

The planetary boundaries framework is a risk assessment for our world. It provides a powerful demonstration of how much we are taking from the Earth and how human activity poses a threat to the habitability of our world. The 2009 paper ended on an optimistic note, stating that as long as the thresholds are not crossed, humanity has the freedom to pursue long-term social and economic development. But in an update of this approach published in

*Science Advances* in late 2023, it was revealed that no fewer than six boundaries (climate change, biodiversity loss, land use change, biogeochemical flows, fresh water use, and chemical pollution and plastics) had been exceeded. The authors exhorted readers to see this 'as a renewed wake-up call to humankind that Earth is in danger of leaving its Holocene-like state'.

Some of the ideas that underpin the planetary boundaries approach can be traced to the Gaia hypothesis, conceived in the 1960s by the British chemist, inventor and maverick environmentalist James Lovelock (1919–2022) and refined in the 1970s in close collaboration with American evolutionary biologist Lynn Margulis (1938–2011), when both worked for NASA. Gaia is the ancient Greek goddess of Earth, and the Gaia hypothesis is concerned with the *self-regulation* of planet Earth by living things and how the living world interacts with the physical world on a grand scale to maintain the habitability of our planet. Lovelock viewed the Earth as a fully integrated entity that was adept, over geological timescales, at keeping conditions within certain boundaries favourable to life. One of the examples used by Lovelock is how Earth's temperature has stayed within a range that allows liquid water to be abundant at the planet's surface since the earliest times, despite marked changes in the input of solar radiation.

A key feature of the Gaia hypothesis is *homeostasis* – the ability of a system to function and maintain stable internal conditions despite external changes. A good example of homeostasis is how the human body regulates its temperature. When you get too hot, you sweat to cool down; when you are too cold, you shiver to generate heat. This self-regulation keeps the body within a temperature range that is favourable for survival. Lovelock's Gaia hypothesis saw planetary-scale homeostasis as a key feature of Earth history.

Many geoscientists were irritated by the Gaia hypothesis and the idea of Earth behaving like a superorganism (not that Lovelock ever claimed the biosphere was conscious or that it anticipated environmental change). Gaia was embraced by some fringe green groups but its mystical connotations were problematic with mainstream science. The evolutionary biologist Richard Dawkins (b. 1941) was especially critical of Gaia, emphasising that natural selection operates on individual genes, not the entire planet. Dawkins saw the hypothesis as teleological (suggesting Earth has a purpose) and incompatible with Darwinian natural selection. He did acknowledge that Earth's biosphere influences its climate and ecosystems, but maintained that local adaptations and feedback mechanisms are fundamental drivers of change rather than a planetary-scale self-regulation system.

Earth science does not regard the Earth as a superorganism, but study of its interconnected planetary-scale systems and the processes involved in the natural long-term self-regulation of Earth's climate have become fundamental concerns. Gaia helped to shape the idea of Earth as a self-regulating system where large-scale feedbacks are an integral part of maintaining planetary habitability (Chapter 17).

While the Gaia hypothesis was an easy target for criticism, it played a role in encouraging others to view Earth as a complex, interconnected system rather than as separate parts (atmosphere, biosphere, lithosphere, hydrosphere, cryosphere) and to think about long-term planetary-scale interactions with feedbacks between biological and physical processes. In key respects the holistic planetary-scale thinking of the Gaia hypothesis and the debates it stimulated set down at least some of the intellectual foundations for the emergence of what is now called *Earth system science*.

Earth system science views the Earth as an interconnected system, exploring planetary-scale interactions between the lithosphere (solid Earth), atmosphere (air), hydrosphere (water), biosphere (life) and cryosphere (ice). It is concerned with understanding the planetary-scale cycling of elements such as carbon and nitrogen and how they are influenced by human activity. Since the Second World War, geology has evolved from a discipline concerned with the description of rocks and landscapes at the local and regional scales to trying to understand how the entire planet works and why it changes. A planetary-scale picture has come to the fore because it helps us to understand the complex relationships between natural processes and human activities, allowing us to address global challenges like climate change, resource deple-tion, biodiversity loss and pollution. All of these can influence how our planet functions.

The planetary boundaries framework offers several vital lessons for humanity. Earth's systems are finite, and exceeding thresholds can lead to irreversible damage. It's safer, cheaper and more effec-tive to stay within these boundaries than to fix problems after they escalate. We understand the problems and we have solutions – we need more evidence-based decision-making, global cooperation and long-term thinking. The question remains whether these can be implemented in time.

# CHAPTER 39

# The Sixth Extinction

Hedgehogs were a common sight when I was a young boy, but I haven't seen one in my south Manchester garden for over ten years. In October 2024, the International Union for Conservation of Nature (IUCN) classified the hedgehog (*Erinaceus europaeus*) as 'near threatened' and highlighted the urgency of conservation efforts. The hedgehog population in the UK countryside has plummeted by about 75 per cent since 2000, a remarkably rapid decline caused by habitat loss, pesticide use and predation. The loss of hedgerows, field margins and natural green spaces has left hedgehogs without safe shelter. Is the hedgehog destined to join the dodo and the dinosaurs? Will it be a casualty of the Sixth Extinction?

Earth history is punctuated by intervals that saw catastrophic losses of life when plants and animals were not able to adapt quickly enough to changes in their environment. The fossil record shows five big extinctions over the past 500 million years, each caused by a planetary-scale trauma. These mass extinction spikes stand out above a long-term background of steady species loss caused by climate shifts, competition, habitat changes and evolutionary processes. During a mass extinction, it's not only that the rate of extinction accelerates rapidly; the rate of new species creation (speciation) can't keep up, so death literally overruns life.

Our planet has seen an exceptionally rapid loss of biodiversity since the beginning of the industrial period. All indicators point to a sixth mass extinction being under way. The wave of species losses we have witnessed in the modern era is quite different to earlier mass extinctions because it is the result of the work of just one species: humans. Earth is now in the midst of a biodiversity and extinction crisis on a scale not seen since the dinosaurs were wiped out 66 million years ago. This time *we* are the planetary-scale trauma driving the extinction.

While the roots of the Sixth Extinction extend back thousands of years to the demise of the ice age megafauna in the Upper Pleistocene, the most dramatic rates of human-driven species loss have taken place during the age of global exploration and especially during the industrial period of the last two centuries. Most famously the flightless dodo (*Raphus cucullatus*) that was endemic to the island of Mauritius in the Indian Ocean is an icon of extinction caused by the arrival of humans. When Dutch sailors reached Mauritius in 1598 they brought rats and pigs who disturbed the natural ecosystem and gorged on dodo eggs. Even though its flesh was tough, the dodo was eaten by sailors – it was not frightened of humans and easily caught. There is some uncertainty about the last recorded sighting, but the dodo was extinct by 1681, less than a century after its first encounter with us.

The most rapid human-caused extinction befell the magnificent Steller's sea cow (*Hydrodamalis gigas*) that grazed on the shallow water kelp beds around the Commander and Aleutian Islands in the southern part of the Bering Sea. It was described in 1741 by the German-born naturalist Georg Wilhelm Steller (1709–1746). Steller's sea cow was a whale-sized beast reaching 10 metres in length. It was curious, often approaching boats, and then was easily harpooned. It may have already been in decline because its range had been more extensive around the North Pacific in the

Pleistocene, but humans struck the final blow. Fur traders seeking sea otter pelts hunted this beast to extinction because Europeans wanted fancy hats. It provided a convenient source of fresh meat during hunting expeditions – there were even reports its thick blubber tasted of almonds. This docile giant was hunted, beached and butchered to extinction within just twenty-seven years of it being formally described.

The thylacine (*Thylacinus cynocephalus*) is a classic example of an animal that was perceived as a threat to human interests and eradicated by colonial settlers. This carnivorous marsupial was native to Australia and New Guinea and commonly known as the Tasmanian tiger because of the dark stripes along its back. About the same size as a Labrador, it preyed on birds and small mammals. This animal went extinct on mainland Australia after the introduction of the dingo about 3,500 years ago but survived into the twentieth century on Tasmania. European settlers viewed the thylacine as a threat to livestock, so the government offered bounties from 1888 to 1909, leading to mass killings. Large numbers were shot, trapped and poisoned. The last known thylacine died in Hobart Zoo on 7 September 1936, marking the end of a lineage that had existed since the Oligocene Epoch some 25 million years ago. There have been reported sightings in the wild in the decades since, but none confirmed. The thylacine was finally declared extinct by the IUCN in 1982.

The passenger pigeon (*Ectopistes migratorius*) was once the most abundant bird in North America, with some 3 to 5 billion individuals. Passenger pigeon DNA shows they had been present in North America in vast numbers since the end of the ice age. Such were their numbers that a flock could take days to pass overhead. European settlers developed a taste for these birds, and they quickly became a popular source of cheap meat in the cities of the East Coast. It seemed inconceivable in the middle of the nineteenth century that such an enormous population could be exterminated by human activity, but the passenger pigeon proved no match for mass forest clearance and European hunters. They were caught in nets, clubbed to death and shot from the sky in vast numbers. In

1878 a single hunter in Michigan killed 3 million birds. In just a few decades, the passenger pigeon was brought to the brink of extinction by mass hunting and habitat destruction – large areas of its nesting grounds were destroyed by woodland clearance. The last known specimen, Martha, died in Cincinnati Zoo on 1 September 1914.

The period after 1950 is often called the Great Acceleration, representing the most intense phase of human influence on Earth's systems. This is when biodiversity loss shifted gear because of habitat destruction for agriculture and urban development, climate change and the overexploitation of Earth's natural resources. All the iconic extinction stories above took place *before* the Great Acceleration. Plant and animal species are now disappearing at somewhere between a hundred and a thousand times the natural background extinction rate (Chapter 24).

The media focus is commonly on animals we have lost and especially on charismatic animals such as rhinos and orangutans at risk of extinction, but plants face many threats too. Flowering plants have shown a remarkable ability to adapt to changing environments. They were among the big winners after the extinction at the end of the Cretaceous Period, to the extent that most of today's terrestrial ecosystems are completely dependent upon them. Botanists have warned that 45 per cent of the world's known flowering plants could be threatened by extinction.

There is still a huge amount of work to be done recording species of plants and fungi, with biodiversity dark spots in Southeast Asia and Oceania where plants have not been systematically recorded. We have only scraped the surface with fungi. There may be 2.5 million species of fungi on Earth, and we have barely recorded 10 per cent of them. It is difficult to keep pace because thousands of new plant species are named each year. The struggle to get to grips with present-day biodiversity underscores how the fossil record presents an incomplete picture of past diversity, given the bias in the record of ancient ecosystems, where fossil bones and hard shells are more easily preserved than petals and leaves.

Many plant species that are extinct in the wild can been seen in botanical gardens. The mission of the Royal Botanic Gardens, Kew, is to 'understand and protect plants and fungi for the well-being of people and the future of all life on Earth'. One of the rarities that can be seen at Kew is *Brighamia insignis*, a slender-stemmed plant with long glossy green leaves and striking, trumpet-shaped flowers. Often called 'cabbage on a stick', it once grew on the steep cliffs and rocky crevices of Hawaii. The last individual was recorded in the wild in the summer of 2012, where it was once pollinated exclusively by a hawk moth that is now extinct. Ironically, while *Brighamia insignis* is extinct in its natural habitat, it is a popular ornamental house plant that is easy to grow. Whether in a specialist botanical greenhouse or on a window sill, each living plant is part of its conservation. Unless it can be reintroduced into the wild, its survival is completely dependent on humans.

There are many examples of successful conservation projects where species have been brought back from the brink of extinction. The red kite (*Milvus milvus*) is a magnificent bird of prey with black-tipped red wings and a long, reddish-brown forked tail. It was regarded as a pest in the UK and hunted to extinction in most parts of the country by the early twentieth century. After a successful reintroduction programme, which involved bringing chicks from Spain via British Airways in 1990, it has recovered dramatically in just three decades from a handful of breeding pairs in west Wales. Red kites are now a common sight across Wales and in most English counties.

Before the 1920s, wolves were an important native predator in Yellowstone National Park, but they were completely wiped out by 1926 due to hunting, poisoning and government predator control programmes. This had profound impacts on the ecosystem. With the apex predator removed, elk populations exploded, and their browsing led to the decline of various tree species, including willow and aspen. Fewer trees and shrubs meant a diminished habitat for songbirds, less dam-building material for beavers, and an overall decline in wetlands and biodiversity. The reintroduction of wolves in the 1990s had a spectacular impact, producing a cascade of ecological changes. It helped restore balance to Yellowstone's ecosystem, benefiting many species and even altering the physical landscape. Elk numbers declined and they avoided open ground, allowing willow and aspen to grow along rivers and stabilise the banks. Beavers built more dams, increasing wetland diversity. Coyote numbers decreased, allowing small mammals like foxes and rodents to flourish.

There is strength in diversity. When crops cannot cope with changing conditions, such as decreased rainfall and warmer temperatures, other varieties can take their place and may even thrive under the changed conditions. But this depends on the availability of a large number of crop varieties. We face major challenges maintaining food supplies for a global population exceeding 8 billion people because we have come to rely on an increasingly small number of crop varieties. The genetic diversity of the nineteenth century no longer exists. A small number of mass-produced crop varieties have served us very well, but when a shift in climate or a new pest threatens their viability, we face a big problem. Having a wide variety of crops protects us from these threats.

And so we are storing seeds for future generations. The Global Seed Vault in Svalbard is a specially designed ice-encased storage unit in a mountainside just outside the capital Longyearbyen. The seed vault came into operation in 2008 as an Arctic sanctuary to safeguard the genetic diversity of crops to help meet the UN goal of eliminating hunger by 2030. The vault stores several hundred million seeds from thousands of varieties of plants from over eighty countries.

It can be easy to overlook the many invaluable functions – such as pollination by insects and delivering clean water and clean air – that ecosystems perform for humanity. These are known as *ecosystem services*. When natural habitats are degraded they lose their capacity to provide these vital services. This threatens food security because crops rely on healthy ecosystems for pollination, fertile soils and pest control. Without a diversity of species, these systems break down. The biodiversity crisis of the modern era is not just about the loss of individual species; it is also about the loss of abundance – what Stanford biologist Rodolfo Dirzo (b. 1951) has called *defaunation* in the case of animals. Habitat destruction threatens the interconnected web of life – from earthworms to rainforests – that supports Earth's ecosystems and human well-being.

The final part of Elizabeth Kolbert's 2014 book *The Sixth Extinction* includes a quote from the American biologist Paul Ehrlich (b. 1932): 'In pushing other species to extinction, humanity is busy sawing off the limb on which it perches.' In other words, by killing everything else off, we're putting our own species at risk of dying off with them. The fossil record shows that life recovered and biodiversity increased after each mass extinction, but this happened over many millions of years, and we do not have that luxury. We can take heart from the success of conservation and rewilding efforts around the world, but to halt the global biodiversity decline we need to make rapid and radical change.

# CHAPTER 40

# The Anthropocene: A Human Planet

In February 2000 at a global environmental change conference in Mexico, one of the delegates became so frustrated that he interrupted a speaker with a comment that would change how we think about our relationship with planet Earth. Paul Crutzen had spent much of his distinguished career investigating the impact of humans on the environment. In 1995 he shared the Nobel Prize in Chemistry for his work on the formation and decomposition of ozone in the atmosphere. His increasing frustration with the talks at the Mexico conference led him to stand up and declare: 'Stop using the word Holocene. We are not in the Holocene anymore. *We are in the Anthropocene!*'

Crutzen's intervention sparked a movement to formally define a new geological epoch. This idea gathered momentum later that year, after Crutzen penned a short article with the American biologist Eugene Stoermer (1934–2012), who had coined the term Anthropocene in the 1980s. Their article set out some preliminary ideas around components of the Earth system where humans were becoming a dominant force. They cited how carbon dioxide emissions from fossil-fuel burning far exceeded those of natural emissions. They described how up to 50 per cent of Earth's land surface had been transformed by human action and how the global nitrogen cycle had been radically altered by the extensive use of fertilisers in agriculture.

The Holocene began 11,700 years ago and is the shortest epoch in the geological record. It has seen the rise of agriculture, permanent settlements, complex urban civilisations and the Industrial Revolution. If most of the Holocene bears the imprint of humanity, what made Crutzen declare it was over? Why do we need a new geological epoch? Crutzen and others have argued that human activity, which now rivals some of the great forces of Nature in its impact on the functioning of the Earth system, has shifted Earth far away from the stable conditions of the Holocene. The Holocene has been changed so profoundly and irreversibly by human action that the Holocene label is now redundant. Humans have so ruptured Earth history that it was time to write a new chapter.

You can't just add a new epoch to the geological timescale; there is a formal process, with rules, protocols and hurdles to clear. The geological timescale is governed by a body known as the International Commission on Stratigraphy (ICS). This is a group of geoscientists who make decisions on the names, boundaries and ages for the eras, periods and epochs that form our Earth history. The ICS works to ensure consistency in the use of these terms and how geologists define and correlate Earth's history worldwide. They are the guardians of the geological timescale.

Boundaries between periods are typically defined using clear markers in the rock record that reflect major environmental, biological or geochemical changes in Earth's history. A geological boundary must be recognised globally. In 2008 the beginning of the Holocene was formally recognised by the ICS by an abrupt climate warming at the end of the last glacial period. This climate change is defined with unusual precision in a Greenland ice core at a depth of 1,492.45 metres, which is dated to 11,700 years before the year 2000. This warming can be seen in geological archives around the world, including lake and marine sediment records and in stalactites in limestone caves.

A new geological epoch must also have a specific location where the boundary between epochs is clearly recorded, and this boundary must be visible in rocks around the world. A reference section of rock layers (strata) that serves as the official standard for defining a geological time unit is called a *stratotype*. The boundary must mark long-term, irreversible changes rather than short-lived events.

In 2009 the Anthropocene Working Group (AWG) was set up by the ICS to explore whether the Earth had entered a new geological epoch defined by the impact of human activity on the planet's systems. One of the group's main tasks was to determine whether the term had any geological significance, and it went on to seek a formal starting point for the Anthropocene and decide on the best indicator of global human impact preserved in the sediment record. Would the AWG choose coal dust, radioactive fallout from nuclear weapons testing, concrete, microplastics or chicken bones? The start of the Anthropocene remains a topic of lively debate, and several dates have been put forward. The AWG spent fifteen years working on these matters.

So when did the Anthropocene begin? Paul Crutzen initially suggested it began with the invention of the steam engine in the late eighteenth century and the onset of carbon emissions from the burning of coal in the early industrial era. The ice cores from Greenland and Antarctica record the beginning of the Industrial Revolution via increased concentrations of carbon dioxide gas trapped in frozen air bubbles.

An early-onset Anthropocene was proposed by William Ruddiman (b. 1943) at the University of Virginia. He has argued that the development of farming and large-scale forest clearance led to an increase in greenhouse gases about 8,000 years ago that can be seen in the ice core records. Others have argued that human populations in the early Holocene were too small to have had a major global impact. The early Anthropocene hypothesis has fallen out of favour because it pales into insignificance compared to the global impacts of the industrial era.

The Anthropocene concept is not about tracing the earliest example of human impact on the environment; it is about recognising when humans became a *dominant* player in the functioning of the planet. The Anthropocene did not begin when humans first started to tinker with Holocene ecosystems – that milestone has been lost in the noise of human prehistory. The Anthropocene proper began when humanity began to push planetary boundaries to crisis point.

The Great Acceleration refers to the exponential increase in human impacts since about 1950. It signals a period of unprecedented economic, technological and environmental change associated with a shift in the scale of resource exploitation, pollution, urban development and biodiversity loss that has profoundly impacted the modern world. The Great Acceleration idea was popularised in a 2004 paper published in *Science*, co-authored by Will Steffen (1947–2023), Paul Crutzen and John McNeill (b. 1954). It is closely tied to discussions of the Anthropocene and planetary boundaries since it is during this period that humans became a major influence upon the Earth system.

Fundamental properties like the composition of the atmosphere, the rate of greenhouse warming and rates of species extinction departed abruptly in the middle of the twentieth century from the stability they had shown for the previous 11,000 years. The burning of fossil fuels produces more than 100 times the $CO_2$ of all the world's volcanoes combined. The rate of warming caused by human activity far exceeds natural Holocene climate fluctuations. Rates of extinction now far exceed natural background levels. Deforestation, habitat destruction and pollution are reshaping ecosystems at an unprecedented scale, pushing the Earth into a new state.

A major feature of the Great Acceleration is the boom in dam construction after the Second World War, which is having a profound impact on the world's rivers. Reservoir construction has modified the global water cycle by holding massive volumes of water on the continents. Reservoirs also trap sediments that would normally be transported to the ocean. Recent decades have seen a worldwide fall in river sediment supply to the oceans of some

50 per cent. The Three Gorges Dam on the Yangtze river in China is the world's largest hydroelectric power station. Before its completion in 2006 the Yangtze river transported about 400 million tonnes of sediment to the coast each year. Today there are more than fifty large dams on the Yangtze and its sediment load has fallen by some 75 per cent. There are many other examples from the Rio Paraná and Amazon in South America, the Nile and Zambezi in Africa, the Mekong, Indus and Ganges–Brahmaputra in Asia. This disruption in the transport of water and sediment from the continents to the oceans is a major modification of Earth's natural rock cycle and a prime example of human activity radically transforming surface processes in the Anthropocene across the globe.

In 2023 the AWG submitted its formal proposal to change the geological timescale. It recommended 1950 as the starting point, with the fallout layer produced by the onset of nuclear weapons testing proposed as a suitable global marker horizon for the beginning of the Anthropocene. In June 2023 the sedimentary sequence in a tiny lake in southern Ontario, surrounded by the Great Lakes, was chosen as the stratotype for the Anthropocene. Crawford Lake has a detailed sedimentary record, with annual layering that includes multiple markers of human impact, particularly from around 1950. These include plutonium fallout from nuclear bomb tests, heavy metals and other pollutants linked to the industrial era, as well as microplastics providing evidence of synthetic materials accumulating in the environment. These signatures of human impact can be seen at sites around the world.

On 24 March 2024 the ICS released a statement confirming that it had rejected the proposal for an Anthropocene Epoch as a formal unit of the geological timescale. Until the ICS decides otherwise, we are still living in the Holocene. The proposal was rejected for several reasons. Many did not see the need for change, arguing that the Holocene already included major human impacts. There was disagreement over whether a suitable geological marker existed that signalled the onset of the Anthropocene – the so-called golden spike – and whether a single site could fully capture the global impact of human activity. The fact that earlier studies had proposed different start dates for the Anthropocene did not help the cause. Many members of the ICS found it hard to approve a geological epoch that was shorter than a human lifetime. The ICS statement conceded that the Anthropocene was a term that would continue to be used, and 'It will remain an invaluable descriptor of human impact on the Earth system.'

Members of the AWG were bitterly disappointed by this decision. A heated debate followed, with many seeing the ICS as having missed an opportunity to heighten public awareness and policy urgency around the threats to the living world. The symbolism of humans living in a new geological epoch defined by their actions may have influenced international negotiations on climate action, marine pollution, conservation and sustainable development. The ICS was happy for the term Anthropocene to be designated as 'an event', for which there is no need for a fixed starting date, and which continues to this day.

The survival of humans through the current global environmental crisis depends on how effectively we address the challenges of climate change, biodiversity loss, resource depletion and pollution. Global population is expected to reach 9 billion by 2037, and while human extinction is highly unlikely in the near future, civilisation as we know it could be severely disrupted if these problems are not tackled urgently.

*Homo sapiens* means 'wise human', reflecting the idea that our species is distinguished by intelligence, reasoning and self-awareness. The Anthropocene is a geological interval of our own making, and we have tried-and-tested solutions to tackle the global environmental emergency. Time will tell if we have the wisdom to do so. There is hope that the very idea of the Anthropocene will help to drive action to conserve and restore the living world. Paul Crutzen used to end his own presentations with a photograph of himself and his grandson. This was his clarion call to the audience to preserve the Earth for future generations.

# Index